FEED-FORWARD NEURAL NETWORKS

Vector Decomposition Analysis, Modelling and Analog Implementation

THE KLUWER INTERNATIONAL SERIES IN ENGINEERING AND COMPUTER SCIENCE

ANALOG CIRCUITS AND SIGNAL PROCESSING

Consulting Editor

Mohammed Ismail

Ohio State University

Related Titles:

FREQUENCY COMPENSATION TECHNIQUES LOW-POWER OPERATIONAL AMPLIFIERS, Ruud Easchauzier, Johan Huijsing
ISBN: 0-7923-9565-4
ANALOG SIGNAL GENERATION FOR BIST OF MIXED-SIGNAL INTEGRATED CIRCUITS, Gordon W. Roberts, Albert K. Lu
ISBN: 0-7923-9564-6
INTEGRATED FIBER-OPTIC RECEIVERS, Aaron Buchwald, Kenneth W. Martin
ISBN: 0-7923-9549-2
MODELING WITH AN ANALOG HARDWARE DESCRIPTION LANGUAGE, H. Alan Mantooth, Mike Fiegenbaum
ISBN: 0-7923-9516-6
LOW-VOLTAGE CMOS OPERATIONAL AMPLIFIERS: *Theory, Design and Implementation*, Satoshi Sakurai, Mohammed Ismail
ISBN: 0-7923-9507-7
ANALYSIS AND SYNTHESIS OF MOS TRANSLINEAR CIRCUITS, Remco J. Wiegerink
ISBN: 0-7923-9390-2
COMPUTER-AIDED DESIGN OF ANALOG CIRCUITS AND SYSTEMS, L. Richard Carley, Ronald S. Gyurcsik
ISBN: 0-7923-9351-1
HIGH-PERFORMANCE CMOS CONTINUOUS-TIME FILTERS, José Silva-Martínez, Michiel Steyaert, Willy Sansen
ISBN: 0-7923-9339-2
SYMBOLIC ANALYSIS OF ANALOG CIRCUITS: Techniques and Applications, Lawrence P. Huelsman, Georges G. E. Gielen
ISBN: 0-7923-9324-4
DESIGN OF LOW-VOLTAGE BIPOLAR OPERATIONAL AMPLIFIERS, M. Jeroen Fonderie, Johan H. Huijsing
ISBN: 0-7923-9317-1
STATISTICAL MODELING FOR COMPUTER-AIDED DESIGN OF MOS VLSI CIRCUITS, Christopher Michael, Mohammed Ismail
ISBN: 0-7923-9299-X
SELECTIVE LINEAR-PHASE SWITCHED-CAPACITOR AND DIGITAL FILTERS, Hussein Baher
ISBN: 0-7923-9298-1
ANALOG CMOS FILTERS FOR VERY HIGH FREQUENCIES, Bram Nauta
ISBN: 0-7923-9272-8
ANALOG VLSI NEURAL NETWORKS, Yoshiyasu Takefuji
ISBN: 0-7923-9273-6
ANALOG VLSI IMPLEMENTATION OF NEURAL NETWORKS, Carver A. Mead, Mohammed Ismail
ISBN: 0-7923-9049-7
AN INTRODUCTION TO ANALOG VLSI DESIGN AUTOMATION, Mohammed Ismail, José Franca
ISBN: 0-7923-9071-7
INTRODUCTION TO THE DESIGN OF TRANSCONDUCTOR-CAPACITOR FILTERS, Jaime Kardontchik
ISBN: 0-7923-9195-0
VLSI DESIGN OF NEURAL NETWORKS, Ulrich Ramacher, Ulrich Ruckert
ISBN: 0-7923-9127-6
LOW-NOISE WIDE-BAND AMPLIFIERS IN BIPOLAR AND CMOS TECHNOLOGIES, Z. Y. Chang, Willy Sansen
ISBN: 0-7923-9096-2
ANALOG INTEGRATED CIRCUITS FOR COMMUNICATIONS: Principles, Simulation and Design, Donald O. Pederson, Kartikeya Mayaram
ISBN: 0-7923-9089-X

FEED-FORWARD NEURAL NETWORKS

Vector Decomposition Analysis, Modelling and Analog Implementation

by

Anne-Johan Annema
MESA Research Institute
University of Twente

SPRINGER-SCIENCE+BUSINESS MEDIA, LLC

ISBN 978-1-4613-5990-6 ISBN 978-1-4615-2337-6 (eBook)
DOI 10.1007/978-1-4615-2337-6

Library of Congress Cataloging-in-Publication Data

A C.I.P. Catalogue record for this book is available
from the Library of Congress.

Printed on acid-free paper.

Contents

Foreword

Artificial neural networks experienced the attention of many researchers in neuroscience and computer science during the last decade. This is usually considered as the revival of neural network research after two decades in which the interest in perceptron-based architectures was lost. The little effort in neural networks in the period 1970 to 1985 can partly be explained by the lack of sufficient data processing power in those days, which prevented researchers to demonstrate the computational capabilities of neural nets. Another reason may be found in the booming progress in VLSI design and realisation, which did attract a lot of attention of the research community.

Meanwhile renewed interest for biologically inspired neural computation focused the attention to the powerful capabilities of this approach for parallel processing. After 1985, the success and progress in VLSI realisation provided the prerequisits for a return to research in neural network realisation. Nowadays the advanced VLSI technology allows the realisation of very cost-effective processors and huge memories which can be used to simulate or emulate parallel neural processing.

Many successful neural network applications have been reported. Several neural network architectures have been investigated by comparing and evaluating simulations on adaptivity and performance. A lot of expert knowledge is gained by experience and analyses of learning systems. This is attended with an exploding number of papers, conference contributions and books on neural networks.

In such circumstances one should have a very good reason when publishing yet another book about neural networks. Fortunately, Anne Johan Annema has such reason. Thanks to his different point of view that originated from the wish to obtain specifications for analog hardware modules, he developed the Vector Decomposition Analysis for feed-forward neural networks with back-propagation learning. In this books he explains the analysis method and illustrates its power.

The Vector Decomposition Method appears applicable for the analysis of feed-forward neural nets, whether they are hardware analog or digital or even software implemented. The key to success is the particular choice for the basis for the Vector Decomposition Analysis, which makes the analytical expressions easy to read and easy to handle. Looking back to the very challenging period during which Anne Johan Annema was with the MESA Research Institute at the University of Twente, I realise that his particular choice for the basis of the Vector Decomposition Analysis offered the key to a great deal of demystification of the neural network learning behaviour. This work, which is a comprehensive cultivation of the PhD thesis has evolved into an interesting and tractable book.

Prof.dr. Hans Wallinga

Acknowledgments

This book is a slightly modified version of my Ph.D. thesis. It describes the results of a research project that was carried out at the MESA Research Institute at the University of Twente, Enschede, The Netherlands. The work has been supported by the Foundation for Fundamental Research on Matter (FOM) and the Netherlands Technology Foundation (STW). A number of people contributed to a nice working atmosphere at MESA or/and to the research project. At the risk of forgetting some persons, I'd like to thank:

- Prof. Hans Wallinga, Klaas Hoen, and Remco Wiegerink for many fruitful discussions and for their comments on the manuscript. I very much liked working together with them.

- My (ex-)colleagues Remco Wiegerink, Eric Klumperink, Peter Masa, Ton Ikkink, Clemens Mensink, Henk de Vries, Roel Wassenaar, Cor Bakker, Jan Hovius, Karel Lippe, Han Speek, Ning, and Jan Niehof for a lot of discussion about all kinds of everything.

- Albert Bos of the Applied Chemistry department of the University of Twente, Stan Gielen of the University of Nijmegen and Peter Johannesma for discussions in the field of neural networks.

- Last but not least, I would like to thank my father, who (among others) made it possible to study at a university, and who always is willing to accompany me when I plan to visit car races or when I have to check if everything is all right under the hood of my car.

Anne-Johan Annema

Chapter 1

Introduction

1.1 Neural networks

Neural networks are systems that typically consist of a large number of simple processing unit, called *neurons*. A neuron has generally a high-dimensional input vector and one single output signal; this output signal is usually a non-linear function of the input vector and a weight vector. The function to be performed on the input vectors is hence defined by the non-linear function and the weight vector of the neuron.

This weight vector is adjusted in a training phase by using a large set of examples and a learning rule. The learning rule adapts the weight of all neurons in a neural network in order to *learn* an underlying relation in the training examples. It may be clear that this type of finding a function to be performed by a system is completely *different from programming* a function.

Historical overview

Since the early 1940's, research has been done in the field of artificial neural networks. This early neural network research was motivated by the observation that the biological brain can perform very complicated tasks and yet consists of simple processing elements: the neurons. To mimic the vast computational power of biological brains, attempts were made to model neurons or their behavior [25],[52]. In the 1950's the research on artificial neural networks grew quickly and several neural network models such as the perceptron [60],[61] and the related Adaline [76],[77] emerged. Both models are single neurons with a two-state output signal. The weights of these neurons could be trained by a large number of learning algorithms, of which the "*Hebb's rule*" and the "*Delta Rule*" are the most widely known. All of these learning algorithms adapted the weights of neurons in at most one layer [39],[53],[78].

The large research effort on perceptron-like neural nets continued until the late 1960's. In the previous two decades of significant neural network research, the perceptions could not be trained to perform complicated classification tasks; only (approximately) linear separable training sets such as for example broom-balancing [79] or simple speech-recognition [71] could be learnt successfully.

In 1969, the book entitled "*Perceptrons*" by Minksy and Papert [53] described clearly the shortcomings of single layer perceptrons. It was furthermore speculated by Minsky and Papert that extending the single-layer perceptron towards multi-layer structures would be useless. In part because of the speculations in "*Perceptrons*" and in part because of the failure of perceptrons to be successfully applied to more complex training sets, the research in the field of artificial neural networks almost stopped around 1970.

However, some research on artificial neural networks remained. In 1974, Werbos [80] replaced the binary non-linear function by a differentiable non-linear function (a sigmoid) and discovered a learning algorithm that could be used to train all neurons in a multi-layer perceptron with sigmoids. It was shown that multi-layer nets could successfully be applied to tasks where the previous perceptrons failed. This work remained however almost unknown and did therefore not lead to renewed interest in artificial neural nets. In 1982, Parker also included sigmoids in the original perceptron and rediscovered the weight adaptation rule that had been discovered by Werbos [81]. This (re)discovery did not attract attention nor lead to the revival of neural nets.

Around the same time that Parker re-invented the perceptrons with sigmoids, a number of other neural network models were discovered. The most important models are the Hopfield network [31], the Bi-directional Associative Memory [43], the Boltzmann Machine [28], the self-organizing feature map [40] and the Cellular Neural Network [14].

The interest in perceptron-type neural networks revived with the discovery of the so-called *back-propagation algorithm* in 1986 by Rumelhart, Hinton and Williams [62]. This back-propagation algorithm is practically the same as the algorithms invented by Werbos in 1974 and Parker in 1982 and also uses sigmoids instead of the two-state non-linear function.

This book is dedicated to the analysis, modification and hardware implementation of perceptron-type neural networks that learn according to the back-propagation algorithm or one of the many derivatives of back-propagation. The last chapter of this book also discusses some recent alternative learning algorithms for hardware implemented perceptron-like neural networks (for example weight perturbation).

When and why using neural networks

This section presents some discussions about the *when* and *why* of using neural networks. It is shown that neural networks are merely a new way of solving problems, which way can sucessfully be followed for a number of problems. For some other problems, it is however not useful to use a neural approach.

As mentioned before, a neural network is basically a method to solve a problem. Conventional methods to solving problems are based on modelling the problem and then analytically derive an algorithm that implements a solution to the problem. The main difference between using neural networks and using a conventional method to solve a problem is that:

- neural networks are *trained* to perform satisfactory: in a training phase, training examples are presented to the networks and the weights of the neural network are adapted by a learning rule (see also other sections in this chapter).
- conventional methods typically *use an (analytical or empirical) model* of the task. Using this model, an algorithm is derived that is suffiecient for solving the problem.

The implementation of the solution (neural or conventional) can be either in software or in hardware. Figure 1.1 shows the two ways to implement a solution to a specific problem. It follows that a neural way to implement a solution to a problems is an alternative way of the conventional, algorithmic, way to solve a problem.

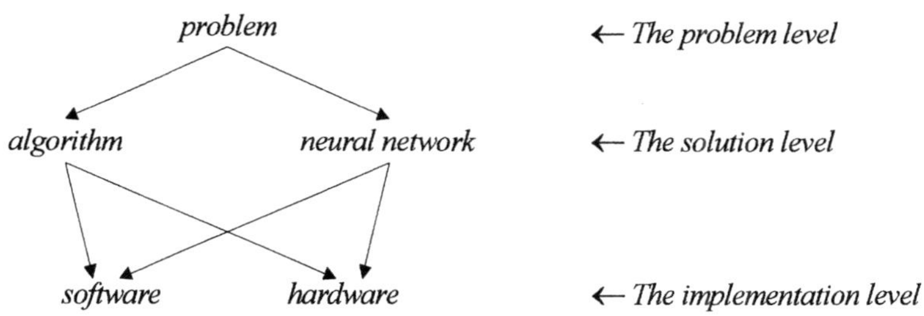

Fig. 1.1 Two ways to solve a problem: neural and algorithmic, and their implementations

For a number of problems, the implementation of the solution in neural networks is useful, while for other problems the solution should not use neural networks. The next part of this section presents a short overview of functions for which implementation in neural networks is useful.

Usefull functions for neural networks

Useful functions to be performed by neural networks can be subdivided into a few categories, which are distinguished by the nature of the problem:

- it is useful to apply neural networks on problems for which no direct algorithmic solution exists, but for which problems examples of the desired responses are available. These examples are required for the training of the neural network. An example of such a problem is the prediction of the water content of cheese in cheese factories [9].
- it is useful to apply neural networks for problems that change over time. The adaptability of the neural network will then be used to adapt the implemented solution whenever the problem changes. Examples for this type of problem are the control of a dynamic and ageing process, or classification of input signals using ageing sensors.
- it is useful to apply neural networks to problems for which only too complicated algorithms can be derived. In this context, "too complicated" means that implemented (conventional) algorithms are either too large, too slow, or consume too much power. Note that in general being "too large" is equivalent to being "too slow". For example in high energy particle colliders, algorithms exist to reduce the data flow, but these algorithms are too slow to be sucessfully applied. Neural networks are therefore used to do high-speed data reduction [24],[51]. A problems for which a too large algorithmic solution can be derived (from a power consumption point of view) is for example dangerous arrhythmia detection in pace-makers [15].

It is *not* useful to train a neural network on problems for which the solution can easily be implemented in an algorithm. If a simple algorithmic solution can be derived, a neural network can also learn this solution, but the neural implementation will generally be larger and less accurate than the direct algorithmic implementation of the solution.

Goal of the book

The goal of this book is twofold. Firstly, the book presents a novel method for the mathematical analysis of neural networks. Using this new method, among others the learning behavior of neural networks can easily be analyzed and specifications for building blocks of neural networks can easily be obtained. Throughout this book, this new method will be used to analyze the learning behavior and derive specifications for feed-forward (perceptron-like) neural networks. The method itself can however also be used for other types of neural networks.

Secondly, the book is dedicated to the design (starting with the derivation of specification and ending with a hardware implementation) of analog hardwired feed-forward neural networks with on-chip back-propagation learning. The *feed-forward* structure has been selected because of its relatively large number of applications. The *on-chip learning capability* is required for applications in which fixed weight configurations cannot be used because of either continuously changing of the controlled process or because of ageing of the neural network itself. Furthermore, for elimination of most mismatches and parameter tolerances that occur in hardwired neural network chips, on-chip adaptation appears to be useful.

The analyses and discussions in this book are dedicated towards a *fully analog implementation* because it is expected that fully analog neural networks have several advantages over other implementations: the required chip area is expected to be small, the power consumption is expected to be low and the speed of operation can be very high compared to other implementation methods.

It has been shown by a number of authors that in implementing the forward part of a feed-forward neural network for moderate speed operations, only problems may occur for the chip area required for the synapses and for the interconnections. However, if also on-chip weight adaptation circuitry is implemented, major problems are expected during training. Therefore, the major part of the analyses in this book are dedicated towards the *learning behavior of feed-forward neural networks* and the *requirements on building blocks* during training.

Section 1.2 gives a course introduction into the feed-forward neural networks. After this introduction on feed-forward nets, section 1.3 presents the basics of the widely used back-propagation rule and section 1.4 gives a short introduction on realizations of feed-forward neural nets. The last section of this chapter presents the outline of this book.

1.2 Feed-Forward Networks

In this section, a course introduction into the feed-forward neural networks is given. This introduction presents the basic architecture of feed-forward neural networks and discusses some basic properties of the architecture.

1.2.1 Architecture of feed-forward neural networks

The basic building block in feed-forward neural networks is the neuron. One single neuron has an N-dimensional input vector and has one single output signal. A neural network consisting of only one neuron is a feed-forward neural network as the output signal is not fed back to the input of the neuron. The schematic of one neuron is shown in figure 1.2.

The single neuron has as its response typically a non-linear function of the weighted sum of all input signals. In this book, the constant bias signal which must be applied to all neurons in a feed-forward neural network is assumed to be included in the input vector $\underline{U}$ of the neuron. Identically, all weights that are used to weight the individual input signals are grouped in a weight vector $\underline{W}$. The response of the single neuron is then

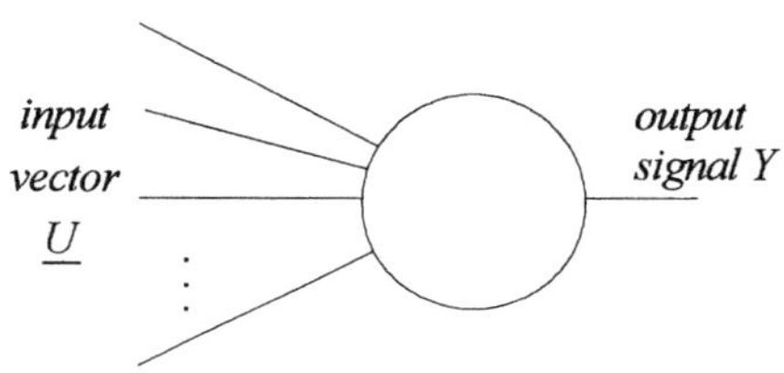

Fig. 1.2 One single neuron

$$Y = f\left(\sum_{n=1}^{N} W_n U_n\right) = f(\underline{W} \cdot \underline{U})$$

The activation function $f(\cdot)$ can have a very large number of shapes. Widely used activation functions are however the sigmoid function, the step function and the piece-wise linear function. The transfer function of these activation functions are given by:

$$f_{sigmoid}(x) = \frac{1}{1+\exp(-a \cdot x)}$$

$$f_{step}(x) = \begin{cases} 1 & x \geq 0 \\ 0 & x < 0 \end{cases}$$

$$f_{pwl}(x) = \begin{cases} 1 & x \geq X \\ \frac{1}{2} + \frac{x}{2X} & |x| < X \\ 0 & x \leq -X \end{cases}$$

In feed-forward neural networks, the neurons are usually grouped in layers, each layer consisting of at least one neuron. The input vector of the neurons in the first layer consist of all externally applied signals and of one constant bias signal.

For *single-layer neural networks*, the output signals of the neurons in the first layer are the output signals of the network. Figure 1.3 shows a typical single-layer feed-forward neural network.

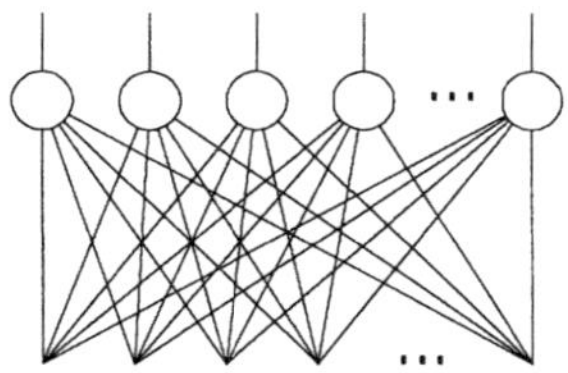

Fig. 1.3 *A general single-layer feed-forward neural network*

Multi-layer feed-forward neural networks consist of multiple layers of neurons. The first layer again consists of a number of neurons, having (N–1) externally applied input signals and one bias input signal. Other layers in the multi-layer network receive their inputs only from all neurons in the previous layer and from one bias signal source. The output signals of multi-layer feed-forward neural nets are the output signals of the neurons in the last layer.

Figure 1.4 shows a *two-layer feed-forward neural network* with five neurons in the first layer and having two neurons in the second layer.

The input vector of the neural network is the input vector of the neurons in the first layer. In this input vector of the network, there are (N–1) externally applied signals and one bias input signal. The input vector of the neurons in the second layer consists of all outputs of the neurons in the first layer and of one bias input signal; the bias signal is represented by the "dangling" inputs. Because the number of output signals of a feed-forward neural network is identical to the number of neurons in the last layer of the network, the net in figure 1.4 has a two-dimensional output vector $\underline{Y}$.

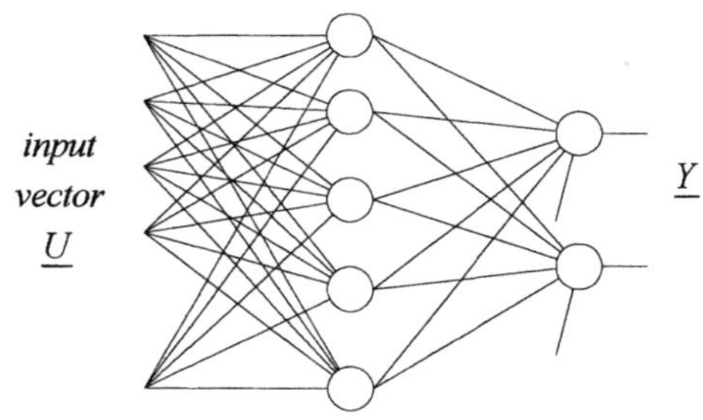

Fig. 1.4 *A two-layer feed-forward neural network*

Three-layer networks and networks with even a larger number of layers can easily be made by adding layers to the two-layer neural network. A general three-layer feed-forward neural network is shown in figure 1.5.

All layers consist of an arbitrary number of neurons. The input vector of the network consists of all externally applied signals and a constant bias input. The input vectors for non-first layer neurons contain all output signals of the neurons in the previous layer and a constant bias input signal.

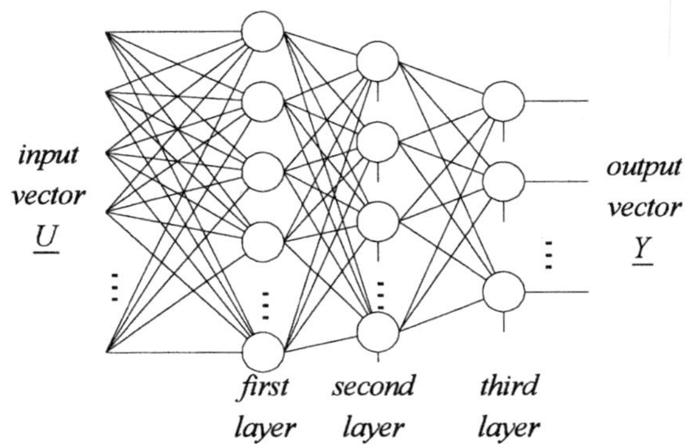

Fig. 1.5 A general three-layer feed-forward neural network

In this book, the analyses and discussions are primarily concerned with single-layer and a special type of two-layer feed-forward neural networks: *the single-output two-layer neural network.* However, a number of analyses and results in this book can be extended to general multi-layer neural networks.

The class of *single-output two-layer feed-forward neural networks* is a sub-class of the general two-layer feed-forward neural networks; the restriction is that there be only one neuron in the second (output) layer. This restriction to single-output two-layer nets has been made for two reasons:

- it simplifies the mathematical analyses considerably as a large part of unnecessary interaction between neurons in a neural network is omitted, and
- subdivision of an *M*-output two-layer neural network into *M* single-output two-layer neural networks is efficient from a neural network learning point of view [3].

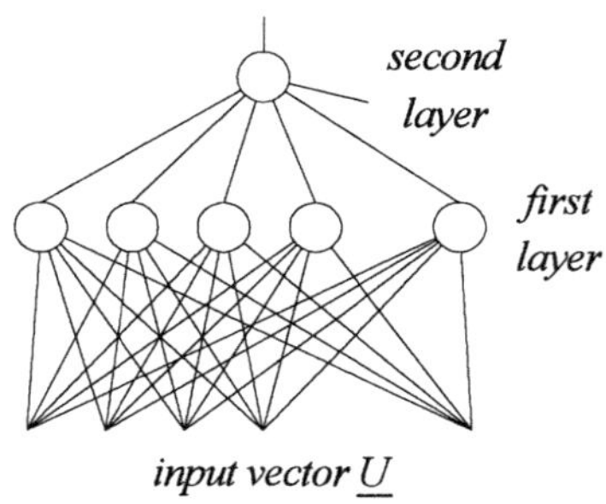

Fig. 1.6 Single-output two-layer feed-forward neural network structure

Figure 1.6 shows a general single-output two-layer feed-forward neural network. This network structure has an arbitrary number of neurons in the first layer, and an arbitrary number of input signals, but has only one neuron in the output layer and hence only one output signal.

1.2.2 Applications for feed-forward neural networks

Feed-forward neural networks can be used to perform several tasks. In literature a large number of tasks are published, including ECG abnormality detection, data reduction, speech recognition, balancing tasks, plant control, sensor signal processing and so on. Very roughly, these tasks can be subdivided into just two classes:

- function approximation, and
- pattern classification

For *function approximation* applications, the feed-forward neural network is trained to learn some function from an *N*-dimensional space to some other *M*-dimensional space. Applications can be in the following areas:

- balancing, (of pole-cart systems for example)
- plant control
- robot (-arm) control
- sensor signal processing

Pattern classification is basically an associative application; the feed-forward neural network is trained to specify the class to which input vectors belong. In this book, we will concentrate on pattern classification applications. Applications in this field include:

- ECG abnormality detection to detect dangerous arrhythmia [15]
- data reduction by deletion of irrelevant data for high-energy physics [51],[24]
- speech recognition [47],[64]
- optical data interpretation [63],[16]
- sensor signal processing [9]

In this book, we will assume that feed-forward neural networks are trained to perform a pattern classification task. It is however straight forward to extend the analyses and results towards function approximation tasks. The major difference for the two types of applications is that for *pattern classification* tasks the desired responses (or target values) for training examples are either "high" or "low", while for *function approximation* the desired (or target) responses for examples can have any value inside the output range of the neurons in the last layer of the network.

1.2.3 Capabilities of feed-forward neural networks: some theorems

It had been shown by the end of the 1980's that two-layer feed-forward neural nets can approximate any function and that two-layer neural nets can be used to classify any non-convex part of the input space [5],[32],[35],[69]. These proofs increased the interest of many research groups in feed-forward neural networks, although they are merely existence proofs. In this section, some of the theorems are presented and their implications for feed-forward neural networks are discussed.

Kolmogorov's theorem

The first theorem that we discuss in this section is the existence theorem by Kolmogorov [42], which has been adapted for feed-forward neural networks [26],[27],[36],[68]:

Theorem 1.1: A two-layer neural network with $2N+1$ neurons in the first layer and with suitable transformation of the input signals, can exactly implement any function in an N-dimensional input space (for $N \geq 2$)

This theorem states that there exist a two-layer feed-forward neural network with $(2N+1)$ neurons in the first layer that can arbitrarily accurate approximate a given N-dimensional function if the activation functions of the neurons have some specific shapes (dependent on the function to be approximated) and if the input signals are transformed by a continuous function ψ (dependent on only the dimension of the input space N). However, this theorem just shows the existence of such a neural network; it does not show how to find the weights nor does it show which activation functions are required for the neurons. The functions ψ that transform the input signals can however be found since they are only dependent on the dimension of the input space [68].

Universal approximation theorem

Some other proofs show that two-layer feed-forward neural networks with bounded, continuous and nonconstant (non-polynomial) activations functions, can approximate any function arbitrarily well. Proofs are published among others in [20],[32],[41],[46],[68].

Theorem 1.2: A two-layer feed-forward neural network with appropriately smooth, saturating first-layer activation functions can make an arbitrarily accurate approximate of any function

These proofs show the existence of a two-layer feed-forward neural network that can arbitrarily well approximate any given function and show furthermore that worst-case infinitely many neurons in the first layer are required.

An illustration

A simple illustration of the two theorems in this section is given below. Assume that one whishes to train a feed-forward neural network to approximate the two-dimensional function $g(u_1, u_2) = sin(A \cdot u_1) \cdot sin(A \cdot u_2)$. This function is shown in figure 1.7:

According to *Kolmogorov's theorem*, this two-dimensional function can be implemented with a two-layer feed-forward neural network with $(2N+1) = 5$ neurons in the first layer.

The theorem does not state how to find the activation functions of any of the neurons nor does it give a clue as how to find the weights for the neurons. Only the transformation ψ of the input signals can be calculated because the dimension of the problem is known. However, for this simple example, we can ommit the input-signal transformation function ψ. Furthermore, it is easy to verify that the network in figure 1.8 can exactly implement the function $g(u_1, u_2)$ by noting that:

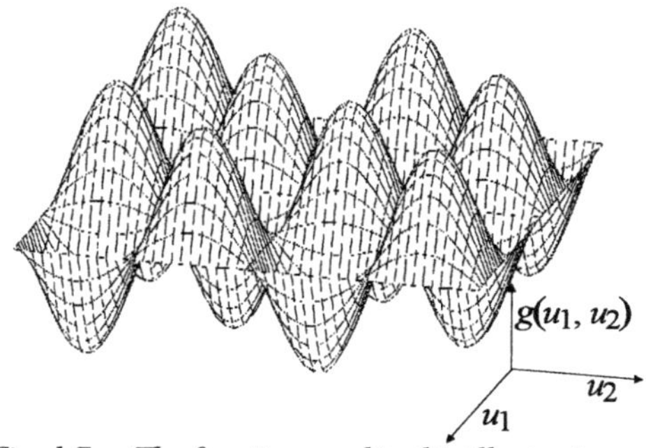

Fig. 1.7 The function used in this illustration: $g(u_1, u_2) = \sin(A \cdot u_1) \cdot \sin(A \cdot u_2)$

$$\frac{-1}{2}\cos(u_1 + u_2) - \frac{1}{2}\cos(u_2 - u_1) = \sin(u_1) \cdot \sin(u_2)$$

Note that to implement this function, only two neurons in the first layer are required. Hence, in the neural network in figure 1.8, three first-layer neurons need not to be used (the associated weights of the neuron in the second layer can be zero). This is however an artifact of using a relatively simple function; to exactly implement *any* 2-dimensional function one needs all five neurons in the first layer and one needs to transform the input signals by functions ψ.

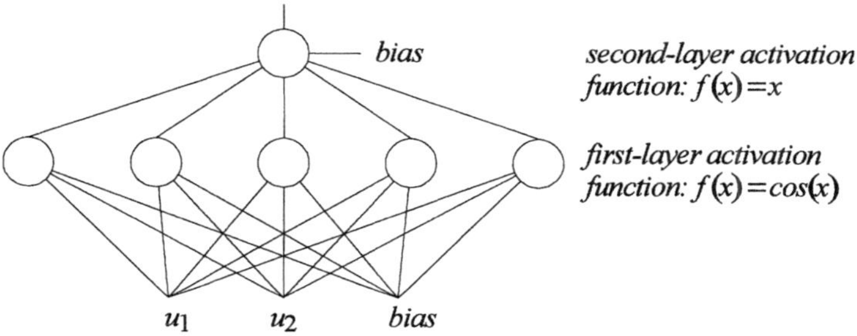

Fig. 1.8 Minimum sized network to perform any function on an 2-dimensional input space; to implement $g(u_1, u_2) = \sin(A \cdot u_1) \cdot \sin(A \cdot u_2)$ *use the specified activation functions*

Theorem 1.2 states that there exists a two-layer neural network (with an arbitrary saturating and non-polynomial activation functions) that can arbitrarily well approximate any function. For the example used in this section, this can easily be illustrated by noting that using $2N$ neurons in the first layer, one can arbitrarily well approximate a block-shaped response or a Kronecker-δ in an N-dimensional input space. An illustration is given in 1.9 for the one-dimensional case:

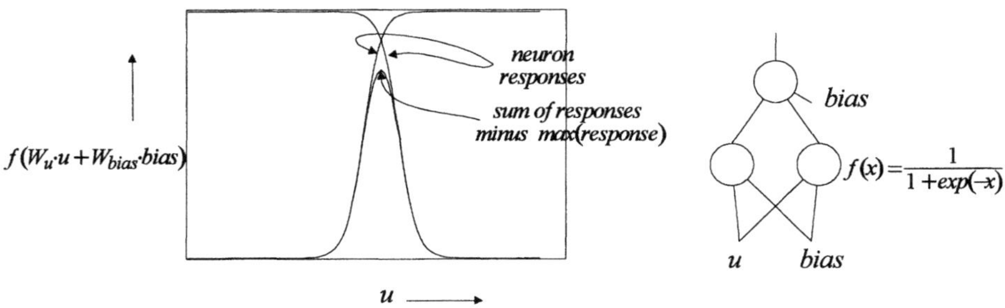

Fig. 1.9 The sum of two sigmoids and a negative constant can arbitrarily well approximate a (Kronecker delta or block) pulse; illustration for one-dimensional case

By adding the responses of two first-layer neurons of a two-layer neural networks and subtracting a constant (to compensate for the average value of the neuron responses), one can obtain a gaussian-like pulse (as shown in figure 1.9). In case of very steep edges of the neuron responses, this gaussian-like shape approximates a block-shape with a finite width.

To obtain a "block-shaped" response in an N-dimensional space, one needs for every dimension two neurons as shown in figure 1.8. The width of the obtained block can be adjusted by properly positioning of the sigmoidal responses. Blocks with infinitely small width are Kronecker-δ-pulses. It is well known that an infinite number of scaled Kronecker-δ pulses can approximate any function.

A one-dimensional illustration of this is given in figure 1.10 for the function $g(u_1,u_2)=sin(A{\cdot}u)\,\sin(A{\cdot}u_2)$.

In the figure, the desired function is approximated by a large number of blocks. Note that the approximation of the sine can also be made using the gaussian-like shapes plotted in figure 1.9.

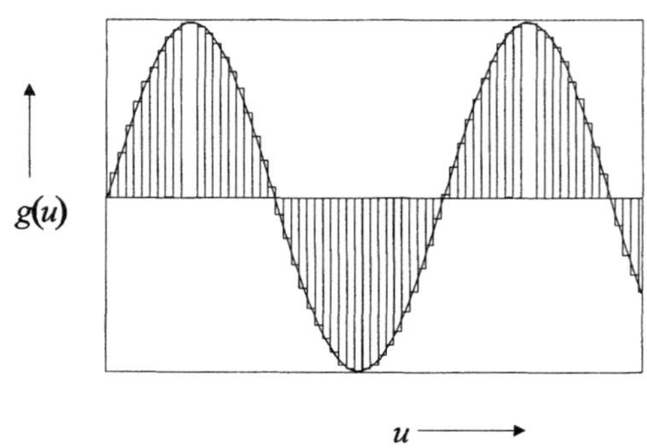

Fig. 1.10 The desired (or target) function and its approximation using blocks

It has been shown that for one-dimensional input signals, block-shaped network responses and gaussian-like shaped network responses can be constructed by using two first-layer neurons for each block or gaussian. As argumented, for N-dimensional input spaces, block-shaped network responses (or gausian-like responses) require $2N$ first-layer neurons. The subtraction or addition of the responses of the first-layer neurons' responses is taken care of by the neuron(s) in the second layer.

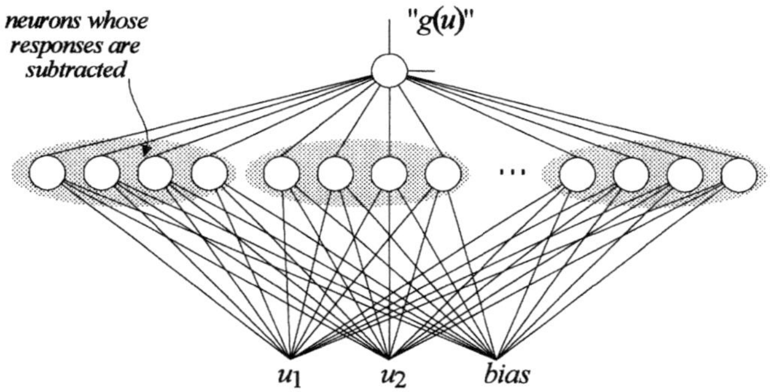

Fig. 1.11 Two-layer neural network to approximate any (here two-dimensional) function

The approximating of any a function using blocks can be improved by using more and more narrow blocks. In the limit, the approximation is made using scaled Kronecker-δ pulses. As mentioned before, it is known that using an infinite number of Kronecker-δ pulses, any function can be implemented exactly.

It follows that two-layer feed-forward neural networks can approximate any function arbitrarily well. However, the number of first-layer neurons required for the approximation increases with the required precision of the approximation and increases with the number of extrema in the function. This latter point can easily be demonstrated by increasing the gain-factor A in the desired function $g(u_1,u_2)=sin(A\cdot u_1)\cdot \sin(A\cdot u_2)$ to infinity: the required number of first-layer neurons in a two-layer net with arbitrary non-polynomial activation functions increases to infinity. Note that for the minimum-sized network in figure 1.8 the required number of first-layer neurons remains two.

Classification capabilities

In 1987, Lippmann [48] stated in one of his papers that a single-layer feed-forward neural network with linear synapses and a sigmoid-like activation function could separate any N-dimensional input space into two parts, separated by a hyperplane[1].

Theorem 1.3: Single neurons (with linear synapses and sigmoid-like activation functions) separate the input space into two parts, separated by a hyperplane.

Training sets that can be classified completely by one hyperplane are denoted *linearly separable training sets*. The classification capabilities of single neurons can then also be described as:

Theorem 1.4: Single neurons (with linear synapses and sigmoid-like activation functions) can only perfectly classify training sets for which the convex[2] hulls of the two classes to be separated are disjoint.

In the paper by Lippmann [48], it was argumented that two-layer feed-forward neural networks could only classify any convex part of the input space. This argument was illustrated by some graphics that show that the neuron(s) in the second layer combine the responses of the neurons in the first layer in an "AND"-type way. In this way, the classifications made by all neurons in the first layers were "AND"-ed together forming a convex space.

In a later paper of Huang and Lippmann [34], it was shown that two-layer feed-forward neural networks can also classify non-convex parts of the input space. This and other papers on the classification capabilities do however not show how to obtain the weights that result in the desired classification.

Theorem 1.5: A two-layer feed-forward neural network (with linear synapses and sigmoid-like activation functions) can classify any (non-)convex part of the input space.

More theoretical results on the classification capabilities of feed-forward neural networks can for example be found in [57],[83].

[1] The hyperplane of a neuron is usually defined as the plane in the input space where the weighted input of the neuron is zero. A more thorough introduction on hyperplanes is provided in section 2.3

[2] A region (or sub-space) is convex if the line that connects any two points that are in the region, only contains points in the region.

Classification performance

In statistics, pattern classification can be performed by a so-called *minimum-error Bayesian classifier*. This type of classifier first calculates the Baysian probability or class conditional probability for an input vector $\underline{U}$. In other words, the Bayesian probability gives the probability that an example $\underline{U}$ belongs to a class C_i:

$$p(C_i \mid \underline{U}).$$

According to the Bayes rule, this class conditional probability is given by

$$p(C_i \mid \underline{U}) = \frac{p(\underline{U} \mid C_i)\ p(C_i)}{p(\underline{U})}. \tag{1.3}$$

where $p(\underline{U} \mid C_i)$ is the conditional probability on an $\underline{U}$ if the class is C_i
$p(C_i)$ is the probability on class C_i
$p(\underline{U})$ is the unconditional probability on $\underline{U}$

In Bayesian classifiers, this probability is calculated for *each* class C_i, and then assigns $\underline{U}$ to the class with the highest probability [58]:

$$\{\underline{U} \in C_i \mid \ p(C_i \mid \underline{U}) > p(C_j \mid \underline{U}),\ i \neq j\}. \tag{1.4}$$

In this section, a number of papers [7],[21],[37],[58],[72] that deal with the correspondence between the Bayesian classifiers and feed-forward neural networks (trained to perform a classification task) are summarized in one theorem. A short discussion of the effect of this theorem is included in this section.

Theorem 1.6: A feed-forward neural network trained to perform vector classification (using the MSE cost function during training) approximates the Bayesian a posteriory estimation on class-probability.

Note that the *a priori* distribution of examples is the distribution of the input vectors (determined by the process that provides the input vectors), whereas the *a posteriori* distribution corresponds to the examples in the training set. For a feed-forward neural network that is trained to classify N-dimensional input vectors into two classes, theorem 1.6 states that (using a learning rule that minimizes the MSE criterion) the function learned by the network approximates the Bayesian estimation of the distribution of the training examples.

It is stated by for example Telfer and Szu [72] that for pattern classification tasks this approximating of the Bayesian estimate is a disadvantage because the functional complexity needed for approximating the a posteriori probabilities is larger than that needed for approximating only class boundaries. In other words, a feed-forward neural network which is trained to approximate the Bayesian a posteriori estimation on class-probability requires more neurons than a feed-forward neural network which is used to only classify input vectors.

1.3 Back-Propagation

It has already been mentioned that the function that is performed by neurons depends on the weight vector of the neurons. These weight vectors are usually determined in a so-called "training-phase" using a learning algorithm. For feed-forward neural networks, the backpropagation rule [62] is probably the most widely used learning algorithm. The backpropagation rule was discovered by Werbos in 1974 [80], rediscovered by Parker in 1982 [81] and again rediscovered by Rumelhart et al. in 1986.

During training of the neuron, subsequently input vectors $\underline{U}^p$ and their associated desired response D^p are presented. For every p^{th} training vector, the back-propagation algorithm adapts the weight vector of all neurons in the neural network in order to minimize a predefined cost function E. This cost function is typically the Mean Squared Error cost function:

$$E = \frac{1}{2P} \sum_{p=1}^{P} \left(D^p - Y^p\right)^2 = \frac{1}{P} \sum_{p=1}^{P} E^p \tag{1.5}$$

The minimization of the cost function is done using gradient descent minimization, resulting in

$$\Delta \underline{W}_{i,j} = -\eta \frac{\partial E}{\partial \underline{W}_{i,j}} = -\frac{\eta}{P} \sum_{p=1}^{P} \frac{\partial E^p}{\partial \underline{W}_{i,j}}, \tag{1.6}$$

where: i *is the index of the layer of the neuron*

j *is the index of the neuron in layer* i

with η a small, positive constant. In a frequently used approximation of (1.6), the weight vector of the neuron is adapted after presentation of any training example. This "*local learning*" or "*on-line*" approximation is allowed for small learning rate η [62], and is used in this book.

Some problems

It is frequently observed that feed-forward neural networks trained with the back-propagation algorithm converge very slowly, and that the minimum in the *energy landscape* (the cost E as a function of all weights for a specific training set) found by the network was likely to be non-global. Many modifications to the back-propagation algorithm were proposed and (almost) all of them were intuitive modifications of the original formulation [1],[2],[4],[11],[12],[44],[82]. This large amount of modified back-propagation algorithms is illustrated by the full name of the recently proposed YPROP algorithm, which is the abbreviation for "Yet another accelerating technique for the back PROPagation" [4].

Other authors report on very different weight adaptation algorithms that result in fast convergence and a reduced likelihood of sticking in local minima [9],[59],[70]. These algorithms generally require information of all training examples and are very hard to implement in hardware. Because of this latter point, no further attention is paid to these algorithms in this book; an overview of some of these algorithms is presented in [22].

1.4 Realizations of feed-forward networks

Because neurons perform relatively simple operations on their input vectors and because neural networks are assumed to be inherently insensitive to analog hardware related problems such as noise, mismatch, offsets, drift and so on, implementing neural networks in analog hardware seems obvious. One of the first attempts to implement neural nets in analog hardware were made in the 1960's by Widrow [77]. In those days, another motivation to realize neural networks in analog hardware was the fact that no powerful (digital) computers were available.
After the revival of neural networks in the mid 1980's, powerful digital computers were available. Because of the ease to program neural nets and their learning algorithms, most neural network research was dedicated to software implementations. However, software implementations have the inherent drawback that the calculations are performed sequentially whereas in neural networks the calculations are in principle performed in parallel. Another disadvantage of software implemented neural networks is that these implementations require a (relatively) large computer. Because of these drawbacks of software implemented neural nets, attempts were made to implement neural networks in hardware.

Ways of hardware implementation
Several ways to implement feed-forward neural networks in hardware are followed. One way is to implement feed-forward neural nets with multiple transputers, every neuron (or a few neurons) is then represented by one transputer [10],[45],[65]. This way of implementing a neural network only distributes the computation over the network but is not an actual hardware implementation because every transputer is in fact a computer and runs its own sequential software. Another way is to implement the neural network in fully digital hardware [38],[67].
For applications that require fast signal processing or very low power consumption, the feed-forward part of the network must be analog. In literature, three types of analog hardwired feed-forward neural networks can be distinguished, based on the way in which the weights are modified. In this section these three approaches are discussed. At the end of this book, a few novel approaches that are suited for on-chip learning neural networks are presented.

- The first type of hardware feed-forward neural network is the non-learning network [50]. For this type of network, the weights are either hardwired through the implementation of fixed gain multipiers. Note that this method is equivalent to using a ROM to store the weights.
  In this case, the weights to be hardwired must be calculated before operation of the neural network chip, which is usually done on a computer which runs a neural network simulation program that includes a model of the actual hardware neural network chip. The eventual level of performance therefore depends heavily on the matching between the model used in the simulations and the actual implemented neural network. In literature, this matching problems is assumed to be a major limitation for this type of network because perfect matching between the simulated model and the implemented model cannot be achieved and because the matching heavily affects the attainable level of performance of the chip.

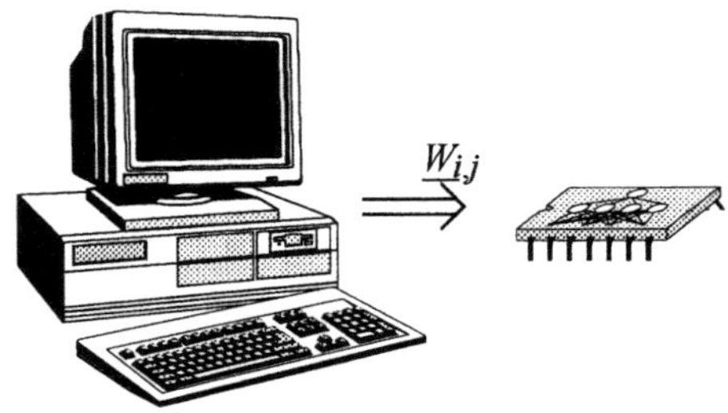

- A second widely used way to implement feed-forward neural networks in analog (or mixed analog-digital) hardware is the construction of a neural network with externally adjustable weights [12],[13],[23],[29],[49],[82]. For these realisations, a host computer is used to train a software model of the neural network. The resulting weights are down-loaded to the chip; the weights are usually stored in a kind of RAM.
  Because of mismatch between the simulation model and the hardware implemented neural network, fine-tuning of the weights is usually required. This fine-tuning of the weights in the chip is done with the so-called "chip-in-the-loop" method. In this "chip-in-the-loop" method, the chip is used to perform the forward pass and the host computer is used to only perform the feed-back (weight adaptation) pass. In this case, the matching between the software and the hardware model is improved considerably with respect to the situation without "chip-in-the-loop" training.

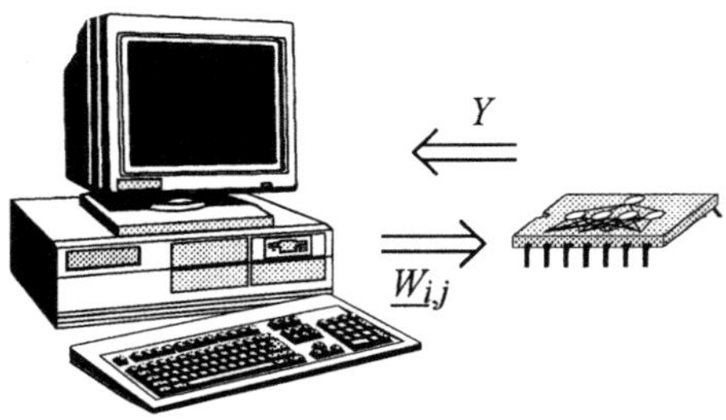

- A third approach to implementing feed-forward neural networks is the implementation of analog *on-chip learning* neural networks [19],[66],[73],[75]. In this case, both the feed-forward structure and all circuitry required to adapt the weights are realized on the chip. Advantages of this approach are among others the high learning speed of the chip (which is caused by the parallel operation and the absence of interface with a host computer) and the small overall size.

  Disadvantages include the high demands on the weight adaptation hardware and the chip area that is used for the weight adaptation circuitry which is only used during weight adaptation. In several publications it is argued that analog on-chip learning neural networks are not feasible, which is motivated by simulations showing that the resolution for the weights *during training* must be about (equivalent) 14 to 16 bits [6],[18],[30], which cannot be reached by analog circuits in a small chip area.

  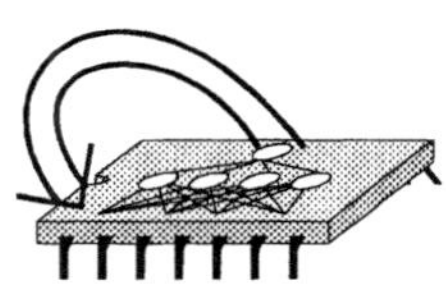

  In these simulations, however, the limited resolution resulted in discretization errors for weights and weight increments, whereas the resolution in analog neural networks is limited by noise and offsets which is fundamentally different. It is shown in [54] that the conclusions drawn from simulations on limited resolution networks *cannot* directly be used for fully analog implementations of on-chip learning neural nets.

Derivation of specifications

It is generally accepted to use simulations to estimate the requirements on building blocks for neural networks [17],[55],[56],[74]. This experimental determination of requirements (typically using a trial-and-error method for specific training sets), is widely accepted because in the mainstream of neural network research, the multi-layer feed-forward neural nets are still viewed as black boxes [8],[33] because of the complex non-linear dynamics of the neural networks.

A major disadvantage of using simulations to estimate the effect of changing some parameter is that *causal* relations do not become clear and that no (or little) insight is obtained. In the second chapter of this book, a mathematical method is introduced which can be used to (among others) derive specifications for building blocks for neural networks. In a number of chapters in this book, the method is used to analyze the learning behavior of feed-forward nets and is used to derive requirements on building blocks. Because of the generic nature of the mathematical method, its application is by no means limited to feed-forward-type neural networks.

1.5 Outline of the book

A novel mathematical analysis method which can be used to analyze the learning behavior of neural networks *and* can be used to derive specifications for building blocks is introduced in chapter 2. In this book, this method is applied to feed-forward neural networks that learning according to the back-propagation algorithm. Chapters 3 to 7 present analyses of effecs in single-layer feed-forward neural networks. The 8th to 11th chapter present analyses of two-layer feed-forward neural nets. Hardware realizations, including design and literature overviews, are presented in chapters 12 and 13.

Single-layer network analyses

Using the novel mathematical analysis method of chapter 2, easy-to-read mathematical expressions are derived for the adaptation of the weight vector of neurons in a single-layer feed-forward network in chapter 3. These equations show how the weight adaptation depends on various properties of the training set and on a number of parameters of the learning algorithm.

In analog hardware implementations of neural networks, it appears to be advantageous to use two-quadrant multipliers instead of four quadrant multipliers because two-quadrant multipliers have a lower circuit complexity, require less chip area and consume less power than four-quadrant multipliers. However, the neurons then need unipolar input signals. The consequences on training time of translating the training set of single neurons in input space, such that unipolar input signals are obtained, are discussed in chapter 4.

In integrated analog neural networks, the input signals are inherently subject to mutual cross-talk via capacitances, inductances and resistive power supply lines. The effects of coupling between input signals of neurons on the learning behavior are analyzed in chapter 5.

In chapter 6, precision requirements for single-layer feed-forward neural networks are derived for analog weight adaptation circuitry. For these circuits, the imprecision is caused by among others leakage, offsets and charge injection of switching transistors. It will be shown that the derived specification can directly be translated in a minimum required size for the weight storage capacitor.

The seventh chapter of this book presents an anlysis of discretized weight adaptations. This analysis can directly be used to estimate the minimum number of bits required for digital implementations *or* can be used to estimate the absolute minimum size for weight storage capacitors in fully analog implementations of on-chip learning neural nets.

Two-layer network analyses

It is frequenty observed that during training of multi-layer feed-forward neural networks, the MSE sticks for a long time at a non-minimum level and then "suddenly" drops to a significantly lower level. Chapter 8 presents a mathematical analysis of the occurrence of this type of temporary minima during training for single-output two-layer neural networks (with backprop learning). The analyses in chapter 8 show that sticking in temporary minima is inherent to multi-layer networks learning. A number of numerical results illustrates the analytical conclusions.

Analyses in chapter 9 show that for two-layer feed-forward neural networks with back-prop learning, the bias signal applied to neurons directly affects the learning behavior of the total neural network. Estimations on (near) optimal biasing neurons are presented. Another problem in two-layer feed-forward neural network training is the encounter of local minima. In chapter 9 an analysis of redundancy-induced encounter of local-minima *and* an improvement for the back-propagation learning algorithm are presented.

In the original formulation of the back-propagation algorithm, a Mean Squared Error cost function is minimized. Chapter 10 discusses the effect on training and on the eventual level of performance of using other types of cost functions.

During training of a multi-layer feed-forward neural network using back-propagation, the derivatives of activation functions must be included in weight adaptations. In chapter 11, a number of analyses and discussions are presented of the effect of mismatch between the implemented approximation of the derivative and the actual derivative. It will be shown that some types of mismatch can reduce requirements on accuracy during learning significantly.

Hardware implementation of feed-forward networks

Chapter 12 presents the design of a neuron which can be used in on-chip learning multi-layer neural networks. Included in this chapter are extensive literature overviews, including circuit schematics and short discussions, of synapses (multipliers) and activation functions.

For on-chip learning neural nets, weight adaptation is required. Chapter 13 presents the design of an analog circuit which can be used in (limited sized) multi-layer feed-forward neural networks. It will be shown that offsets on the weight adaptation signal limit the applicability of analog on-chip learning nets. Solutions and possible solutions are presented in chapter 13 and in the recommendations part of chapter 14.

1.6 References

[1] L.G.Allred and G.E.Kelly, "Supervised Learning Techniques for Backpropagation Networks", in Proc. IJCNN, San Diego CA., vol. 1, pp. 721-728, 1990

[2] R.Aloni-Lavi, Y.Metzger and Y.Stein, "A BP Variant with Improved Convergence and Generalization", in Proc. IJCNN, Baltimore, vol. 1, pp. 932-937, 1992

[3] R.Anand, "Efficient neural network algorithms for multiclass problems", Ph.D. Thesis, Syracuse University, New York, 1992

[4] D.Anguita, M.Pampolini, G.Parodi and R.Zunino, "YPROP: yet another accelerating technique for the back-propagation", in Proc. ICANN 93, Amsterdam, pp. 500, 1993

[5] M.Arai, "Mapping Abilities of Three-Layer Neural Networks", in Proc. IJCNN, Washington DC., vol. 1, pp. 419-423, 1989

[6] K.Asanovic and N.Morgan "Experimental Determination of Precision Requirements for Back-Propagation Training of Artificial Neural Networks", in Proc. 2nd Int. Conf. on Microelectronics for Neural Networks, pp. 9-15, Munich, 1991

[7] E.Barnard, "Comparison of backpropagation and Bayes results for using a priori information in neural net classifiers", IEEE tr. Neural Networks, vol. 4, pp. 794-802, 1993

[8] I.Bellido and E.Fiesler, "Do Backpropagation Trained Neural Networks have Normal Weight Distributions?", in Proc. ICANN 93, Amsterdam, pp. 772-775, 1993

[9] A.Bos, "Artificial Neural Networks as a Tool in Chemometrics", Ph.D.Thesis, Twente University, The Netherlands, 1993

[10] C.F.Chang and B.J.Sheu, "Digital VLSI Multiprocessor Design for Neurocomputers", in Proc. IJCNN, Baltimore, vol. 2, pp. 1-6, 1993

[11] H.W.Chen and V.W.Soo, "An Adaptive Back-Propagation Learning Method: A Preliminary Study for Incremental Neural Networks", in Proc. IJCNN, Baltimore, vol. 1, pp. 713-718, 1992

[12] M.R.Choi and F.M.A.Salam, "Implementation of Feedforward Artificial Neural Nets with Learning using Standard CMOS VLSI Technology", in Proc. IEEE ISCAS, Singapore, pp. 1509-1512, 1991

[13] J.Choi and B.J.Sheu, "VLSI Design of Compact and High-Precision Analog Neural Network Processors", in Proc. IJCNN Maryland, vol. 2, pp. 637-641, 1992

[14] L.O.Chua and L.Yang, "Cellular Neural Networks: Theory", IEEE Trans. Circuits And Systems, vol. 35, pp. 1257-1272, 1988

[15] R.Coggins, M.Jabri, B.Flower and S.Pickard, "Low Power Intracardiac Electrogram Classification unsing Analogue VLSI", in Proc. Fourth Int'l. Conf. on MicroNeuron, pp. 376-382, 1994, Turin

[16] E.Cosatto and H.P.Graf, "NET32K High Speed Image Understanding System", in Proc. Fourth Int'l. Conf. on MicroNeuron, pp. 413-421, 1994, Turin

[17] M.P.Craven, K.M.Curtis and B.R.Hayes-Gill, "Frequency division multiplexing in analogue neural network", Electronic Letters, vol. 27, no. 11, pp. 918-920, 1991

[18] R.C.Frye, E.A.Rietman and C.C.Wong, "Back-Propagation Learning and Nonidealities in Analog Neural Network Hardware", IEEE Trans. Neural Networks, vol. 2, pp. 110-117, 1991

[19] B.Furman, J.White and A.A.Abidi, "CMOS Analog IC Implementing the Back Propagation Algorithm", in Proc. INNS 1988, Boston, pp. 381

[20] A.R.Gallant and H.White, "On Learning the Derivatives of a Unknown Mapping With Multilayer Feedforward Networks", Neural Networks, vol. 5, pp. 129-138, 1992

[21] H.Gish, "A Probabilistic Approach to the Understanding and Training of Neural Network Classifiers", in Proceedings of the Int'l Conf. on Acoustics Speech and Signal Processing, pp. 1361-1364, 1990

[22] C.de Groot and D. Wurtz, "Plain backpropagation and advanced optimization algorithms: A comparative study", Neurocomputing, vol. 6, pp. 153-159, 1992

[23] I.S.Han and K.H.Ahn, "Implementation of Million Connections Neural Hardware with URAN-I", in Proc. ICANN'93, Amsterdam, pp. 1030-1033, 1993

[24] A.A.Handzel, T.Grossmann, E.Domany, S.Tarem and E.Duchovni, "A neural network classifier in experimental particle physics", Int'l Journal of Neural Systems, vol. 4, pp. 95-108, 1993

[25] D.O.Hebb, "The Organisation of Behavior", New York: Wiley, 1949

[26] R.Hecht-Nielsen, "Kolmogorov's Mapping Neural Network Existence Theorem", Proc. IEEE IJCNN, vol. 3, pp.11-13, 1987

[27] R.Hecht-Nielsen, "Theory of the Backpropagation Neural Network", Proc. IEEE IJCNN, vol. 1, pp. 593-605, 1989

[28] G.E.Hinton and T.J.Sejnowski, "Learning and relearning in Boltzmann machines", in *Parallel Distributed Processing*, vol. 1, chapter 7, Eds. D.E.Rumelhart and J.L.McClelland, Cambridge MA.: MIT Press, 1986

[29] M.Holler, S.Tam, H.Castro and R.Benson, "An Electrically Trainable Artificial Neural Network (ETANN) with 10,240 "Floating Gate" Synapses", in Proc. IJCNN, Washington DC., vol. 2, pp. 191-196, 1989

[30] P.W.Hollis, J.S.Harper and J.J.Paulos, "The Effects of Precision Constraints in a Backpropagation Learning Network", Neural Computation, vol. 2, pp. 363-373, 1990

[31] J.J.Hopfield, "Neural networks and physical systems with emergent collective computational abilities", Proc. Natl. Acad. Sci., vol. 79, p. 2554-2558, 1982

[32] K.Hornik, M.Stinchcombe and H.White, "Universal Approximation of an Unknown Mapping and Its Derivatives Using Multilayer Feedforward Networks", Neural Networks, vol. 3, pp. 551-560, 1990

[33] T.Hrycej, "Modular Learning in Neural Networks", New York: Wiley & Sons, 1992

[34] W.Y.Huang and R.P.Lippmann, "Neural Net and Traditional Classifiers", in Proc. Neural Information Processing Systems, pp. 387-396, 1987

[35] B.Irie and S.Miyake, "Capabilities of Three-layered Perceptrons", in Proc. IJCNN, San Diego, vol. 1, pp. 641-648, 1988

[36] H.Katsuura and D.A.Sprecher, "Computational Aspects of Kolmogorov's Superposition Theorem", Neural Networks, vol. 7, pp. 455-462, 1994

[37] F.Kanaya and S.Miyake, "Bayes statistical behavior and valid generalization of pattern classifying neural networks", IEEE tr. Neural Networks, vol. 2, pp. 471-475, 1991

[38] Y.C.Kim and M.A.Shanblatt, "An Implementable Digital Multilayer Neural Network", in Proc. IJCNN, Baltimore, vol. 2, pp. 594-600, 1992

[39] J.S.Koford and G.F.Groner, "The Use of an Adaptive Threshold Element to Design a Linear Optimal Pattern Classifier", IEEE Trans. Information Theory, vol. 12, no. 1, pp. 42-50, 1966

[40] T.Kohonen, "Self-organized formation of topologically correct feature maps", Biolog. Cybernetics, vol. 43, pp. 59-69, 1982

[41] P.Koiran, "On the complexity of approximating mapping using feedforward networks", Neural Networks, vol. 6, pp. 649-653, 1993

[42] A.N.Kolmogorov, "On the Representation of Continuous Functions of Many Variables by Superposition of Continuous Functions of One Variable and Addition", Doklady Akademic Nauk USSR, pp. 953-956, 1957 (Translations American Mathematical Society, vol. 2, pp. 55-59, 1963

[43] B.Kosko, "Adaptive bidirectional associative memories", Appl. Optics, vol. 26, pp. 4947-4960, 1987

[44] R.Kothari, P.Klinkhachorn and R.S.Nutter, "An Accelerated Back Propagation Training Algorithm", in Proc. IJCNN, Singapore, vol. 1, pp. 165-170, 1991

[45] P.Kotilainen, J.Saarinen and K.Kaski, "Mapping of some neural network algorithms to a general purpose parallel computer", in Proc. ICANN 93, Amsterdam, pp. 1085, 1993

[46] M.Leshno, V.Y.Lin, A.Pinkus and S.Schocken, "Multilayer Feedforward Networks With a Nonpolynomial Activation Function Can Approximate Any Function", Neural Networks, vol. 6, pp. 861- 868, 1993

[47] E.Levin, "Multilayer hidden-control neural architecture for modeling signals produced by nonlinear time-varying systems; application to speech and handwriting recognition", IEEE tr. Neural Networks, vol. 4, pp. 109-116, 1993

[48] R.P.Lippmann, "An Introduction into Computing with Neural Nets", IEEE ASSP Mag, vol. 4, pp. 4-22, 1987

[49] J.B.Lont and W.Guggenbuhl, "Analog CMOS Implementation of a Multilayer Perceptron with Nonlinear Synapses", IEEE Trans. Neural Networks, vol. 3, 1992

[50] P.Masa, K.Hoen and H.Wallinga, "20 Million Patterns Per Second Analog CMOS Neural Network Pattern Classifier", in Proc. ECCTD'93, Davos, pp. 497-502, 1993

[51] P.Masa, K.Hoen and H.Wallinga, "High Speed VLSI Neural Network for High-Energy Physics", in Proc. Fourth Int'l. Conf. on MicroNeuron, pp. 422-429, 1994, Turin

[52] W.S.McCulloch and W.Pitts, "A logical calculus of the ideas immanent in nervous activity", Bulletin of Mathematical Biophysics, vol. 5, pp. 115-133, 1943

[53] M.Minsky and S.Papert, "Perceptrons", Cambridge, MA: MIT Press, 1969

[54] A.F.Murray and P.J.Edwards, "Synaptic Weight Noise During Multilayer Perceptron Training: Fault Tolerance and Training Improvements", IEEE Trans. Neural Networks, vol. 4, pp. 722-725, 1993

[55] D.B.Mundie and L.W.Massengill, "Weight Decay and Resolution Effects in Feedforward Artificial Neural Networks", IEEE Trans. Neural Networks, vol. 2, pp. 168-170, 1991

[56] J.A.G.Nijhuis, "An Engineering Approach to Neural System Design", Ph.D. Thesis, Katholieke Universiteit Nijmegen, The Netherlands, 1992

[57] A.Rao, M.R.Walker, L.T.Clark, L.A.Akers and R.O.Grondin, "VLSI Implementation of Neural Classifiers", Neural Computation, vol. 2, pp. 35-43, 1990

[58] M.D.Richard and R.P.Lippmann, "Neural Network Classifiers Estimate Bayesian *a posteriori* Probabilities", Neural Computation, vol. 3, pp. 461-483, 1991

[59] R.A.Rosario and N.Tepedelenlioglu, "A Rapid Multi-Layer Perceptron Training Algorithm", in Proc. IJCNN, Baltimore, vol. 1, pp. 824-829, 1992

[60] F.E.Rosenblatt, "Perceptron Simulation Experiments", in Proc. of the IRE, pp. 301-309, 1960

[61] F.E.Rosenblatt, "Principles of Neurodynamics", New York: Spartan, 1962

[62] D.E.Rumelhart, G.E.Hinton and R.J.Williams, "Learning internal representations by error propagation", in *Parallel Distributed Processing*, vol. 1, chapter 8, eds. D.E.Rumelhart and J.L.McClelland, Cambridge MA.: MIT Press, 1986

[63] E.Sackinger and H.P.Graf, "A System for High-Speed Pattern Recognition and Image Analysis", in Proc. Fourth Int'l. Conf. on MicroNeuron, pp. 364-375, 1994, Turin

[64] P.K.Sankar, "Multilayer perceptron, model for fuzzy classification of patterns; application to speech recognition", IEEE tr. Neural Networks, vol. 3, pp. 683-697, 1992

[65] M.Schwarz, B.J.Hostica, M.Kesper, P.Richert and M.Scholles, "A CMOS-Array-Computer with On-Chip Communication Hardware Developed for Massively Parallel Applications", in Proc. IJCNN, Singapore, vol. 1, pp. 89-94, 1991

[66] T.Shima, T.Kimura, Y.Kamatani, T.Itakura, Y.Fujita and T.Iida, "Neuro Chips with On-Chip Back-Propagation and/or Hebbian Learning", IEEE J. Solid-State Circuits, vol. 27, pp. 1868-1875,1992

[67] L.Spaanenburg, B.Hoefflinger, S.Neusser, J.A.G.Nijhuis and A.Siggelkow, "A Muliplier-less Digital Neural Network", in Proc. 2nd Int. Conf. Microelectronics for Neural Networks, Munich, pp. 291-310, 1991

[68] D.A.Sprecher, "A Universal Mapping for Kolmogorov's Superposition Theorem", Neural Networks, vol. 6, pp. 1089-1093, 1993

[69] M.Stinchcombe and H.White, "Approximating and Learning Unknown Mappings using Multilayer Feedforward Networks with Bounded Weights", in Proc. IJCNN, San Diego, vol. 3, pp. 7-16, 1990

[70] J.Sun, W.I.Grosky and M.H.Hassoun, "A Fast Algorithm for Finding Minima of Error Functions in Layered Neural Networks", in Proc. IJCNN, San Diego, vol. 1, pp. 715-720, 1990

[71] L.R.Talbert, G.F.Groner, J.S.Koford, R.J.Brown, P.R.Low and C.H.Mays, "A Real-Time Adaptive Speech-Recognition System", Stanford University Technical Report 6760-1, 1963

[72] B.A.Telfer and H.H.Szu, "Energy Functions For Minimizing Misclassification Error With Minimum-Complexity Networks", Neural Networks, vol. 7, no. 5, pp. 809-818, 1994

[73] M.Valle, D.D.Caviglia and G.M.Bisio, "An Experimental Analog VLSI Neural Chip with On-Chip Back-Propagation Learning", in Proc. ESSCIRC 1992, Copenhagen, pp. 203-206

[74] A.Von Lehmen, E.G.Peak, P.F.Liao, A.Marrakchi and J.S.Patel, "Factors Influencing Learning by Backpropagation", in Proc. IJCNN, San Diego, vol. 1, pp. 335-341, 1988

[75] Y.Wang, "A Modular Analog CMOS LSI for Feedforward Neural Networks with On-Chip BEP Learning", in Proc. IEEE ISCAS 1993, Chicago, pp. 2744-2747

[76] B.Widrow and M.E.Hoff, "Adaptive Switching Circuits", IRE WESCON Conv. Record, part 4, pp. 96-104, 1960

[77] B.Widrow, "An Adaptive 'Adaline' Neuron Using Chemical 'Memistors'", Stanford University Technical Report no. 1553-2, October 1960

[78] B.Widrow, "Generalization and Information Storage in Networks of Adaline 'Neurons'", in *Self Origizing Systems 1963*, Eds. M.C.Yovitz, G.T.Jacobi and G.D.Goldstein, Washington DC.: Spartan Books, pp. 435-461

[79] B.Widrow, "The original adaptive neural network Broom-Balancer", in Proc. IEEE ISCAS, Philadelphia, pp. 351-357, 1987

[80] P.J.Werbos, "Beyond Regression: New Tools for Prediction and Analysis in the Behavioral Sciences", Ph.D. thesis, Harvard University, 1974

[81] P.J.Werbos, "Backpropagation: Past and Future", in Proc. IJCNN 1988, San Diego, vol. 1, pp. 343-353

[82] Y.Zhang, G.E.Hearn and P.Sen, "A Modified Learning Algorithm for Backpropagation Network", in Proc. ICANN 93, Amsterdam, pp. 478, 1993

[83] P.Zwieterink, "The Complexity of Multi-Layer Perceptrons", Ph.D.Thesis, Technical University of Eindhoven, The Netherlands, 1994

Chapter 2

The Vector Decomposition Method

2.1 Introduction

When implementing neural networks in either software or hardware, one must have specifications for building blocks (either software functions or hardware circuits) in order to make a sufficiently good implementation. In this context, sufficiently good means that the minimum requirements for proper operation must be satisfied and that at the same time the various building blocks must not be over-specified as this is usually expensive (economically or computationally).

specifications for
neural building blocks

too low	*too high*
≡	≡
no proper operation	*unnecessarily expensive*
(at least in worst-case situation)	*(computationally, economically)*

To derive specifications for the various building blocks for neural networks, a number of ways can be followed. We list three possible approaches and give a short discussion of all.

- Firstly, one can use a trial-and-error method in determining the minimum requirements for various building blocks. A normal and widely accepted procedure is to plot the performance of a neural network as a function of some specifications of building blocks for specific training (or test) sets. The building blocks with specifications that are marginally sufficient to satisfy the requirements on performance are then implemented. This method has several major disadvantages:
  (a): Using a trial-and-error method, generally no causal relations between any specification of any building block and the resulting level of performance of the total neural network is obtained. As a result of this, the network and the learning algorithm cannot easily be modified in order to relax demands on building blocks.
  (b): The specifications for building blocks that result from a trial-and-error method can generally not be generalized towards other training (or test) sets. This effectively means that the trial-and-error procedure must be repeated for every new application.

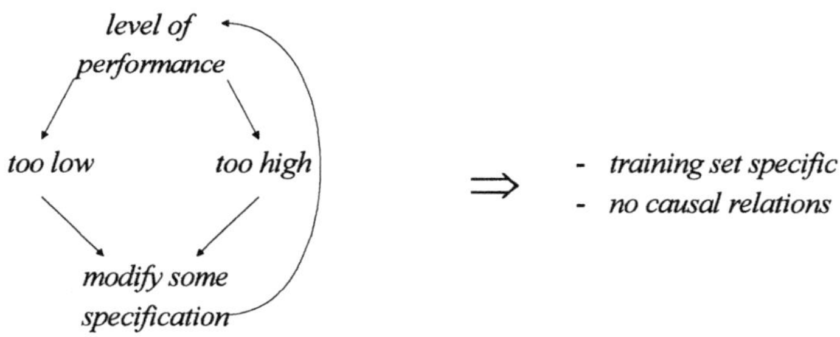

- Secondly, one may analyze the learning behavior or the effects of non-idealities on the operation of a neural network using mathematics. For the learning behavior of neural networks, very complicated expressions result which is due to actually solving strongly non-linear coupled differential equations with stochastic inputs (the training examples). Because of the complexity of the resulting equations, no easy-to-read causal relations can be found nor is insight obtained in the learning process. This lack of insight into neural network dynamics and neural information gathering is the reason why neural networks are usually looked upon as black-boxes.
  However, for the derivation of sensitivities of trained neural networks with respect to non-idealities (such as limited weight precision and noise) conventional mathematical solutions may be useful.

- The third approach is using the *Vector Decomposition Method* (VDM). This VDM is a novel method which can be used to mathematically derive easy-to-read relations for various effects in various types of neural networks. The fundamentals of the method are introduced in this chapter.

Througout this book, the VDM is applied only to feed-forward neural networks with back-propagation learning (or a derivative of the back-prop rule). The VDM is used to analyze the learning behavior of single-layer and multi-layer feed-forward neural networks in (among others) chapters 3, 8 and 9 of this book. Specifications for building blocks and the effects of non-idealities on the learning behavior are analyzed in chapters 4 to 7, 10 and 11 of this book. *However, because of the generic nature of the VDM, the applicability of the VDM is by no means limited to feed-forward neural networks nor to the back-propagation learning rule.* The next section presents the basics of the VDM for general neural networks. The other sections of this chapter present the VDM in more detail and apply the VDM to feed-forward neural networks. Note in all of these sections that applying the VDM to other types of neural networks is straight forward.

2.2 The basics of the VDM

This section presents the basics of the Vector Decomposition Method. The Vector Decomposition Method is based on the assumption that every neuron in a neural network learns something during the training phase. This also includes that after a training phase, every neuron in a neural network did learn something. Note that if for a specific neuron in a neural network the assumptions is not satisfied, it is useless to analyze the behavior of this neuron (or to derive specs) with any method because the neuron is not operating. Hence, the base theorem of the VDM is:

Theorem 2.1: Every neuron in a neural networks learns *something*

With the VDM, this learned "*something*" is now decomposed into two components. One of the components corresponds to the correct part of what has been learnt by the neuron in the neural network and the other part corresponds to the difference between the total "*something*" and the correct part of this "*something*":

Theorem 2.2: *something* $\equiv$ *something good* + *deviation*

In neural networks, the weight vector of a neuron in a neural net represents the "*something*" learned by the neuron while the "*something good*" is determined by the training set. With the VDM, the weight vector and the input vector of every neuron in a neural network is now decomposed into three orthogonal vector components which are related to the "*something good*" for every neuron. This means that a new base is introduced for every neuron in a neural net which base is used to represent the weight and input vector of the corresponding neurons. Note that for a neural network, the number of bases used to decompose weight and input vectors equals the number of neurons in the network.

In the next sub-section of this chapter, the decomposition of the weight vector and the input vector of neurons in a neural network will be discussed in more detail.

2.3 Some notations and definitions

In the previous section of this chapter, it has been mentioned that with the Vector Decomposition Method the weight vector and the input vector of a neuron in a neural network are decomposed into three vector components which are related to "*something good*". In this section, we present the notation for the input and weight vector of neurons and present a few definitions which will be used to define the "*something good*" in the next section.

The notation of input vectors

In most neural literature, the input vectors applied to neurons in feed-forward neural nets are either externally applied to the neural network (for first layer neurons) or applied by all neurons in the previous layer (non-first layer neurons). The constant bias input signal applied to all neurons in the feed-forward neural network is usually not looked upon as an input signal of the neurons.

However, with the VDM the constant bias input signal *is* assumed to be an element of the input vector of neurons. In the notation used in the VDM, all input signals of the neuron in a neural network (including the constant bias input signal) form an N-dimensional input vector, which will be denoted as $\underline{U}$:

$$\underline{U} = (U_1, U_2, \ldots, U_{N-1}, U_{bias})^T. \tag{2.1}$$

To comply with the existing literature on neural networks, the *input space* of neurons in a feed-forward neural networks is defined as the $(N-1)$-dimensional space spanned by the first $(N-1)$ elements of the input vector $\underline{U}$. Henceforth, the input space consists of all non-constant elements of the input vector $\underline{U}$.

The effect of the Nth input signal (with its associated weight) will be explained in the next section of this chapter.

The notation of weight vectors and neuron responses

The weights which are used to weight all elements of the input vector of a neuron form an N-dimensional *weight vector* will will be denoted as

$$\underline{W} = (W_1, W_2, \ldots, W_{N-1}, W_{bias})^T \tag{2.2}$$

The response Y of a neuron is typically a non-linear function $f(\cdot)$ of the sum of the weighted elements of the input vector. This sum is denoted as the *weighted input of a neuron*. With the VDM, the weighted input of a neuron is given by

$$weighted\ input = \sum_{n=1}^{N} W_n U_n = \underline{W} \cdot \underline{U} \tag{2.3}$$

The response Y of a neuron in a neural network is then a specific non-linear function (usually sigmoid-shaped or gaussian-shaped) of the weighted input of the neuron:

$$Y = f(\underline{W} \cdot \underline{U})$$

In this book, it is assumed that sigmoid-like activation functions are used in the neurons. This is however by no means necessary for the VDM-based analyses nor is it necessay for the neural networks as was shown in the introduction on feed-forward neural networks in chapter 1.

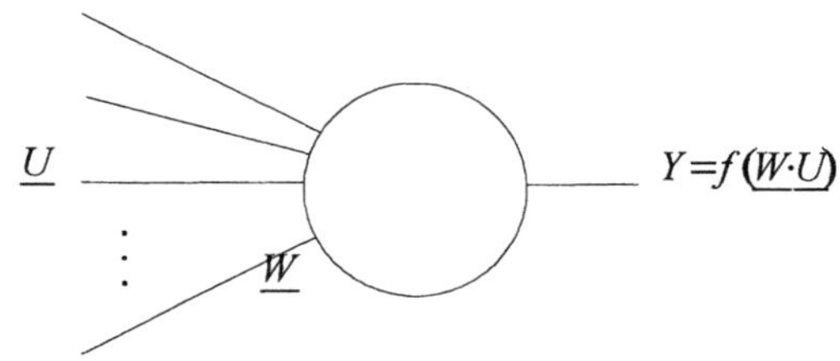

Fig. 2.1 A neuron, and its signals and weights

Figure 2.1 shows one neuron in a feed-forward neural network with its input vector, its weight vector and its response Y. Throughout this book, the notation presented in this sub-section will be used to denote the weight vector, the input vector, the weighted input and the response of the neuron.

The hyperplane of a neuron

For pattern classifications tasks, the sign of the weighted input of a neuron basically determines in which of two classes the present input vector is classified[1]. Hence, the decision boundary at which the decision is made in which class examples are classified is located at examples for which $\underline{W}\cdot\underline{U}=0$. For sigmoid-shaped activation functions and ordinary weight multipliers (i.e. multipliers that perform an ideal multiplication), examples for which $\underline{W}\cdot\underline{U}=0$ form an $(N-1)$-dimensional plane in input space. This (hyper)plane in the input space is given by

$$W_1U_1+W_2U_2+\ldots+W_{N-1}U_{N-1} = -W_NU_N. \tag{2.4}$$

Normally, this $(N-1)$-dimensional plane in the input space is generally referred to as the *hyperplane* that corresponds to a weight vector $\underline{W}$. Note that the vector is given by

$$[W_1, W_2, \ldots, W_{N-1}, 0]^T$$

is perpendicular to the hyperplane. It follows directly from these equations that the Nth input signal (the bias) and the related N^{th} weight take care of the distance between the hyperplane and the origin of input space. This distance given by:

$$d(hyperplane\,,\,origin\_of\_input\_space) = \frac{|W_N\,U_N|}{|[W_1,\,W_2\,,\ldots,W_{N-1},\,0]^T|} \tag{2.5}$$

Definition of *weight vector attractor* and *attractor hyperplane*

During training, the weights of all neuron in a neural network are adapted in order to minimize a predefined cost-function. In this training phase, the neural network will learn to (at least partially) classify the training set. Consequently, the weight vector of neurons in a neural network converge during training towards specific attractors in weight space [1]-[3] corresponding to a (at least local) minimum of the cost function that is minimized during training.

The attractor in weight space for a neuron in a neural net will be called the *weight vector attractor* of a neuron. The weight vector attractor of a neuron will be denoted as $\underline{W}^{ATT}$. This weight vector attractor may be stationary in weight space or may be moving in weight space, as will be shown for feed-forward neural networks in for example chapter 8 of this book.

The hyperplane corresponding to the weight vector attractor will be denoted as the *attractor hyperplane*. This attractor hyperplane is then the hyperplane that results in a (at least locally) optimal performance for the specific neuron in a neural network. In this book, both the terms weight vector attractor and attractor hyperplane will be used, depending on which one results in the most simple explanation or notation.

[1] *For double threshold classification, the degree of confidence of the classification is a monotonic function of the magnitude of the weighted input; a zero weighted input corresponds to likelihood of 0.5 for being in either class. For a more thorough discussion, see chapter 10.*

2.4 The VDM in more detail

In the previous sections of this chapter, the basics of the VDM and a few definition are presented. In this section, we present a discussion of the "*something good*" for a specific neuron in a feed-forward neural network. After this discussion, we show the decomposition of the weight and input vector of the neuron into three vector components.

2.4.1 Decomposition basics

In the VDM, the weight vector and the input vector of neurons are decomposed into three orthogonal vector components. These three vector components are related to the "*something good*" for the neuron. With the definitions in the previous section, the "*something good*" is the direction of the weight vector attractor, or equivalently the direction perpendicular to the attractor hyperplane of the neuron.
The weight vector and the input vector of the neuron are now decomposed into three orthogonal vector components related to the attractor hyperplane:

- one vector component is perpendicular to the attractor hyperplane. This component corresponds therefore to the "*something good*" component in the input and weight vector.
- one vector component is in parallel to the attractor hyperplane of the neuron. This vector component is perpendicular to the "*something good*" component.
- the third vector component is perpendicular to both previous components; this component is related to only the bias input of the neuron.

The next sub-section of this chapter presents the decomposition of the weight vector and the input vector of neurons in a neural network in a mathematical form. This means that the concept of for example "*something good*" is directly translated into a vector component.

2.4.2 The actual vector decomposition

The decomposition of the input vector and the weight vector of a neuron in a neural network is best illustrated in two steps. Note that as the concept of "*something good*" is different for every neuron in a neural network, this decomposistion must be done for every neuron individually. However, it will be shown in for example chapter 8 that symmetry assumptions and clustering assumption will simplify analyses significantly.

The first step is to decompose both the weight and input vector of the neuron into two vectors:

- one vector component is denoted by the superscript *bias* and has elements which are related to only the bias input signal.
- the other vector component is related to all non-bias related elements in the input and weight vectors; these vector components are denoted by the superscript *r*.

The result of this first decomposition step is

$$\underline{U} = \underline{U}^{bias} + \underline{U}^{r} \tag{2.6a}$$

and

$$\underline{W} = \underline{W}^{bias} + \underline{W}^{r}\,. \tag{2.6b}$$

For the vector components in (2.6a) and (2.6b) hold:

$$\underline{U}^{r} \perp \underline{U}^{bias}\,, \tag{2.7a}$$

$$\underline{W}^{r} \perp \underline{W}^{bias}\,, \tag{2.7b}$$

$$\text{and} \quad \underline{U}^{bias} \,//\, \underline{W}^{bias}\,. \tag{2.7c}$$

The non-bias related vector components $\underline{U}^{r}$ and $\underline{W}^{r}$ are now decomposed into two orthogonal vector components, respectively perpendicular and in parallel to the attractor hyperplane. The weight vector and input vector component that are perpendicular to the attractor hyperplane will be denoted $\underline{W}^{h}$ and $\underline{U}^{h}$. These two vector components satisfy

$$\underline{W}^{ATT}\cdot\underline{U}^{r} = \underline{W}^{ATT}\cdot\underline{U}^{h} \tag{2.8a}$$

and

$$\underline{W}^{ATT}\cdot\underline{W}^{r} = \underline{W}^{ATT}\cdot\underline{W}^{h}\,. \tag{2.8b}$$

The other vector components resulting from the decomposition are in parallel to the attractor hyperplane. This vector component is therefore perpendicular to the previous two components. For the input vector, this component is called $\underline{U}^{F}$, with

$$\underline{U}^{F} = \underline{U}^{r} - \underline{U}^{h}\,. \tag{2.9}$$

For the weight vector, this vector component is denoted as $\underline{W}^{\varepsilon}$, with

$$\underline{W}^{\varepsilon} = \underline{W}^{r} - \underline{W}^{h}. \quad (2.10)$$

The input vector and the weight vector of the neuron are decomposed as

$$\underline{U} = \underline{U}^{h} + \underline{U}^{bias} + \underline{U}^{F}, \quad (2.11)$$

$$\underline{W} = \underline{W}^{h} + \underline{W}^{bias} + \underline{W}^{\varepsilon}. \quad (2.12)$$

Using this decomposition introduces a new base for the weight vector and for the input vector of every neuron in a neural network. Note that because in general every neuron in a neural network has a different attractor hyperplane (i.e. attractor for the weight vector in weight space), the decomposition is generally different for every neuron.

2.4.3 Quantification of vector components

Quantification of some of the vector components in (2.11) and (2.12) simplifies the notation in the vector decomposition based analysis method. A simple quantification of the first two components on the right hand side of (2.11) and (2.12) can be made by introducing two unity vectors,

$$\underline{B}^{h} = \frac{\underline{U}^{h}}{|\underline{U}^{h}|} \quad (2.13)$$

and

$$\underline{B}^{bias} = \frac{\underline{U}^{bias}}{|\underline{U}^{bias}|}. \quad (2.14)$$

The norms of the vector components in the $\underline{B}^{h}$ direction are represented by α^{h} for the input vector and by β^{h} for the weight vector. Similarily, the norm of the vector component in the $\underline{B}^{bias}$ direction are represented by α^{bias} for the input vector and by β^{bias} for the weight vector component. Quantification of the $\underline{U}^{F}$ and $\underline{W}^{\varepsilon}$ vector components using unit vectors does generally not simplify the notations and the analyses. With the just introduced notation, we obtain

$$\underline{U} = \alpha^{h}\,\underline{B}^{h} + \alpha^{bias}\,\underline{B}^{bias} + \underline{U}^{F} \quad (2.15)$$

and

$$\underline{W} = \beta^{h}\,\underline{B}^{h} + \beta^{bias}\,\underline{B}^{bias} + \underline{W}^{\varepsilon}. \quad (2.16)$$

The α^{h} and $|\underline{U}^{F}|$ may be associated with the concepts of *relevant information* and *irrelevant information* respectively in the input vectors. Similarly, β^{h} and $|\underline{W}^{\varepsilon}|$ may be associated with *correct knowledge* and *incorrect knowledge* respectively.

2.4.4 An illustration

Figure 2.2 illustrates the decomposition of the weight vector of a neuron for a simple two-dimensional training set (input space). The figure only shows the vector component in the direction of "*something good*": $\underline{B}^h$ and the vector component in parallel to the attractor hyperplane, $\underline{W}^\varepsilon$. The bias vector component is perpendicular to the paper.
The shaded areas in figure 2.2 mark the two classes to be classified by the neuron.

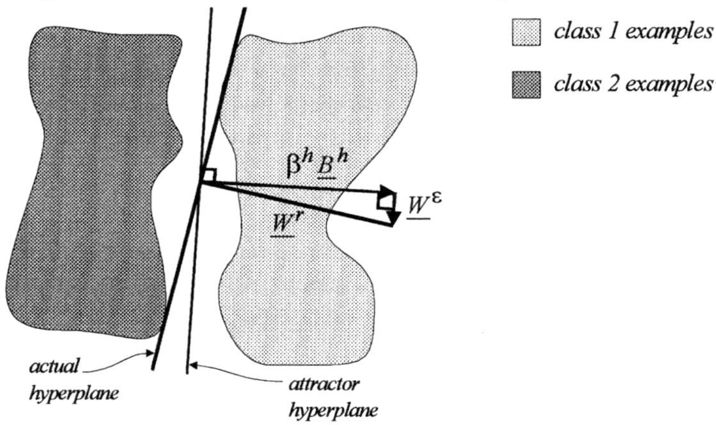

Fig. 2.2 Weight vector decomposition; bias component not shown

As argumented earlier in this chapter, the attractor hyperplane corresponds to the (locally) optimal position of the hyperplane. In general the actual hyperplane does not coincide with the attractor hyperplane, which results in a non-zero norm for the $\underline{W}^\varepsilon$ vector component.

2.4.5 The neuron response

The response of the neuron is a non-linear function of the weighted input of the neuron. With (2.15) and (2.16), the response of the neuron is

$$Y = f(\underline{W}\cdot\underline{U}) = f\left(\alpha^h \beta^h + \alpha^{bias} \beta^{bias} + \underline{W}^\varepsilon \cdot \underline{U}^F\right). \tag{2.17}$$

In the analyses in the next chapters, vector decompositions (2.15) and (2.16) will be used and neuron responses according to (2.17) are used.

2.5 A summary of the VDM

The goal of the Vector Decomposition Method (VDM) is to enable the derivation of easy-to-read expressions that either describe the learning behavior or give specifications for building blocks of neural networks. In this book, the VDM is applied to feed-forward neural networks only; this is however not a restriction of the VDM but merely a choice made by the author.

With the VDM, a new base is introduced for the weight vector and the input vector of every neuron in a neural network. This base is correlated to the optimal weight vector (at a certain point in training *or* after training) for every individual neuron in a neural network. Hence, every neuron in a neural network has its specific base which is used to decompose the weight and input vector of only that neuron.

In figure 2.3 the specific decompositions are indicated by the $\underline{B}^h_{layer,number}$ vectors (the vector perpendicular to the optimal hyperplane) which are specific (and hence generally different) for each neuron in the neural net.

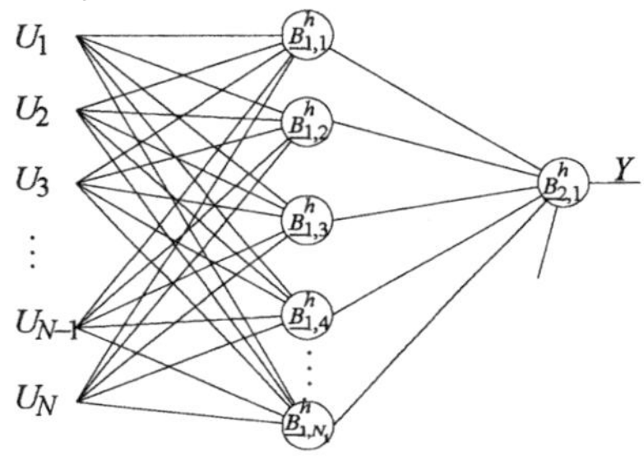

Fig. 2.3 Every neuron in a neural network (here feed-forward with N_1 neurons in the first layer) has its own decomposition based on its specific $\underline{B}^h$.

It appears that by choosing the new base in the way as descibed in this chapter, the learning behavior of neural networks can be described by easy-to-read equations. Relatively simple equations and relations describing learning behavior and the effects of non-idealities on learning are presented for both single and multi-layer neural networks in various chapters in this book.

2.6 References

[1] H.Guo and S.B.Gelfand, "Analysis of Gradient Descent Learning Algorithms for Multilayer Feedforward Neural Networks", *IEEE Trans. Circuits and Systems*, vol. 38, 883-894, 1991

[2] T.M.Heskes and B.Kappen, "Learning processes in neural networks", *Physical Review A*, vol. 44, 2718-2726, 1991

[3] D.B.Parker, Learning Logic, Technical Report 47, MIT, 1985

Chapter 3

Dynamics of Single Layer Nets

3.1 Introduction

A well known sub-class of neural networks is the single-layer feed-forward neural network, in which all neurons are grouped in one layer [3],[4]. A typical single-layer feed-forward neural network is shown in figure 3.1a.

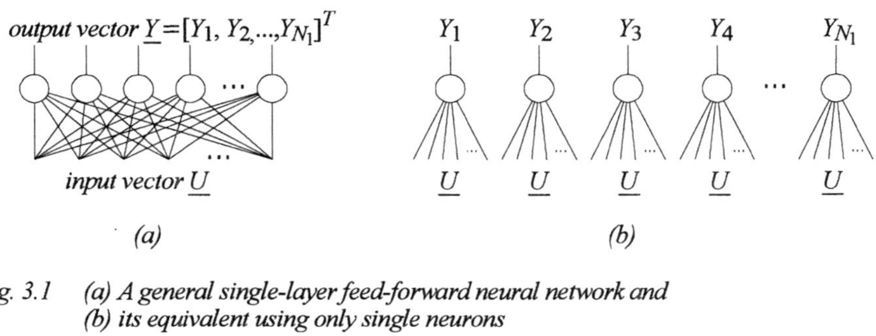

Fig. 3.1 (a) A general single-layer feed-forward neural network and (b) its equivalent using only single neurons

In single-layer feed-forward neural networks, there is no feedback (direct or indirect) from any parameter of the nth neuron to the mth neuron for $n \neq m$. This means that a single-layer feed-forward neural network that consists of N_1 neurons is equivalent to N_1 single neurons that share the same input space, but operate independently.

This sharing of the input space means that the training examples presented to each neuron in figure 3.1b have the same dimension. The training set used to train the neurons is generally different for each neuron. This chapter presents an analysis of the learning behavior of single neurons, which can hence easily be extended towards the analysis of the learning behavior of general single layer feed-forward neural networks.

Training

To perform a classification or approximation task, the weight vector of each neuron has to be determined during a learning (or training) phase. In this book it is assumed that feed-forward neural networks are trained to perform a *classification task.*

In this chapter, it is assumed that the adaptation of the weights is done using the back-propagation learning algorithm. In the training phase, examples and the corresponding desired responses (or target responses) are presented to the neural net. The learning algorithm now adapts the weights of all neurons in order to minimize a predefined cost-function. For the back-propagation rule [6]-[8], the cost-function to be minimized is the Mean Squared Error cost function E:

$$E = \frac{1}{2P}\sum_{p=1}^{P}\left(D^p - Y^p\right)^2 = \frac{1}{P}\sum_{p=1}^{P} E^p \qquad (3.1)$$

where

- p denotes the index of the training example
- P is the total number of examples in the training set
- D^p is the desired (or target) response for the pth training example, and
- Y^p is the response of the network for the pth training example

With the notations introduced in chapter 2, the response of a neuron on an input vector $\underline{U}$ is:

$$Y^p = f(\underline{W}\cdot\underline{U}^p) \qquad (3.2)$$

As stated before, during training the weight vector of the neuron is adapted in such a way that the cost E is minimized. The minimization is done using a gradient descent minimization method, resulting in

$$\Delta\underline{W} = -\eta\frac{\partial E}{\partial\underline{W}} = -\frac{\eta}{P}\sum_{p=1}^{P}\frac{\partial E^p}{\partial\underline{W}}. \qquad (3.3)$$

In (3.3), the factor η is a small, positive constant, usually denoted as the *learning rate*. The use of this factor is to ensure stability during training; i.e. to ensure that the adaptation steps of the weight vector $\underline{W}$ are small enough to ensure that the weight vector converges towards the attractor in weight space (see chapter 2).

[1] Extending the analyses towards function approximation tasks is straight forward as it only means that the desired responses (target responses) are not restricted to "low" or "high", but can have any value inside the output range of the network.

In a frequently used approximation of (3.3), the weight vector of the neuron is adapted after presentation of any training example. This "local learning" or "on-line" approximation is allowed for small learning rates η [7]. We will use this local learning approximation of the original backpropagation in this chapter. With this approximation of (3.3), the adaptation of the weight vector is given by:

$$\Delta \underline{W}^p = -\eta \frac{\partial E^p}{\partial \underline{W}} . \tag{3.4}$$

With (3.2) and (3.4), the adaptation of the weight vector of single neurons is

$$\Delta \underline{W}^p = \eta \left(D^p - f(\underline{W} \cdot \underline{U}^p) \right) f'(\underline{W} \cdot \underline{U}^p) \, \underline{U}^p . \tag{3.5}$$

In this book it is assumed that the feed-forward neural network is used for pattern classification applications. For pattern classification, the desired responses D^p are either "low" or "high". For the training, we use the well known *double-threshold* modification of the back-propagation algorithm [8], which states that there be no adaptation for:

$$Y^p \geq D^p \Big\backslash_{D^p = D^{high}} \quad \text{and} \quad Y^p \leq D^p \Big\backslash_{D^p = D^{low}} \tag{3.6}$$

This means that the response of the network is correct if it is larger than the "high" threshold for patterns that should be classified as "high", *and* the response of the network is assumed correct if it is lower than the "low" threshold for "low" patterns. It has been shown that the double-threshold version of back-propagation performs much better than a strict implementation of (3.5).

A number of analyses in this chapter deal with the learning behavior of single neurons with the double-threshold variant of back-propagation learning. Extensions towards a analyses of the learning behavior using a strict implementation of (3.5) or even when training to perform function approximation tasks is straight forward. A detailed outline of the analyses and discussions in this chapter is given below.

Outline of chapter 3

This chapter presents analyses of the learning behavior of single neurons. As shown in the begin of this section, these analyses are equivalent to analyses of the learning behavior of general single-layer feed-forward neural networks. Section 3.2 presents a few analyses of the learning behavior of single neurons. These analyses are used in sections 3.3 to 3.6 to derive the effects of respectively the learning rate, scaling of input signals, and the bias-input on the learning behavior. With the VDM, relatively simple expressions are obtained for all of these effects.

3.2 Weight vector adaptation with the VDM

In chapter 2 of this book, the Vector Decomposition Method was proposed to be used to analyze the behavior of neural networks. In addition to several publications on the analysis of the learning behavior of feed-forward neural nets [1]-[4],[9],[10], the VDM contributes to further understanding of the behavior of neural networks because easy-to-read expressions are derived as will be shown in this book. In this section, some basics of the VDM are discussed and the VDM is applied to single neurons that learn according to the back-propagation rule.

3.2.1 Weight adaptation of one neuron with the VDM

During the training phase of single neurons, training examples $\underline{U}^p$ and their associated desired (target) responses D^p are presented to the neuron. The weight of the neuron are modified according to the double-threshold variant of the back-propagation algorithm[1] (see the previous section of this chapter).

When training one neuron to classify the examples of a certain training set using the back-propagation algorithm, the weight vector adaptation on the pth training example is given by (3.5). With the VDM, the input vector and weight vector of the neuron are decomposed into three orthogonal vector components as shown in (2.15) and (2.16). The adaptation of the weight vector, $\Delta\underline{W}$, is therefore equal to

$$\Delta\underline{W} \equiv \Delta\beta^h \underline{B}^h + \Delta\beta^{bias}\underline{B}^{bias} + \Delta\underline{W}^{\varepsilon} \tag{3.7}$$

Using the VDM, the three equations that describe the adaptation of the weight vector of the neuron for the pth training example are

$$\begin{cases} \Delta\beta^{h,p} &= \eta\left(D^p - f(\underline{W}\cdot\underline{U}^p)\right) f'(\underline{W}\cdot\underline{U}^p)\, \alpha^{h,p} \\ \Delta\beta^{bias,p} &= \eta\left(D^p - f(\underline{W}\cdot\underline{U}^p)\right) f'(\underline{W}\cdot\underline{U}^p)\, \alpha^{bias,p} \\ \Delta\underline{W}^{\varepsilon,p} &= \eta\left(D^p - f(\underline{W}\cdot\underline{U}^p)\right) f'(\underline{W}\cdot\underline{U}^p)\, \underline{U}^{F,p} \end{cases} \tag{3.8}$$

In this chapter, we will analyze the *average* adaptation of β^h and β^{bias}, the norm of the *average* $\underline{W}^{\varepsilon}$ and the *standard deviation* of $\underline{W}^{\varepsilon}$. In the next part of this section, some general equations for the adaptation of β^h and β^{bias} and for the norm of $\underline{W}^{\varepsilon}$ are presented. The remainder of this chapter deals with the effects on training of various parameters of both the training set and of the learning algorithm.

[1] In the remainder of this book, it is implicitly assumed (unless stated otherwise) that the double-threshold variant of back-propagation is used whenever we refer to the back-propagation algorithm.

3.2.2 Average adaptation of β^h and β^{bias}

The average adaptation of β^h (the *correct knowledge part* of the weight vector, see section 2.4.3) and β^{bias} (the bias-related weight), at any state of the learning process is given by

$$\Delta\beta^h \approx \frac{1}{P}\sum_{p=1}^{P}\Delta\beta^{h,p} \tag{3.9}$$

and

$$\Delta\beta^{bias} \approx \frac{1}{P}\sum_{p=1}^{P}\Delta\beta^{bias,p}. \tag{3.10}$$

where $\Delta\beta^h$ and $\Delta\beta^{bias}$ denote the average adaptations.

Because the adaptation of the weight vector directly depends on the training examples (including their desired responses) and the other parameters in (3.5), it may be evident that the average adaptation of β^h and β^{bias} depend heavily on the training set, on the non-linear function, and on the learning rate η.

3.2.3 Adaptation of $\underline{W}^\varepsilon$

The third vector component of the weight vector, $\underline{W}^\varepsilon$, represents the deviation of the actual weight vector from the locally optimal direction $\underline{B}^h$. Because it may be assumed that learning is non-optimal, the norm of $\underline{W}^\varepsilon$ vector component is generally non-zero. For the derivation of expressions that describe the norm of $\underline{W}^\varepsilon$, it appears to be useful to decompose $\underline{W}^\varepsilon$ into two vector components: one vector component $\overline{\underline{W}^\varepsilon}$ representing the *moving-average* (or short term average), and one component $\underline{W}^{\varepsilon,\nu}$ having *elements with a zero mean value*. With this notation, the $\underline{W}^\varepsilon$ vector component can be written as

$$\underline{W}^\varepsilon = \overline{\underline{W}^\varepsilon} + \underline{W}^{\varepsilon,\nu}. \tag{3.11}$$

In words, the $\overline{\underline{W}^\varepsilon}$ vector represents the moving-average deviation of the actual weight vector from the local optimal direction of the weight vector, and the $\underline{W}^{\varepsilon,\nu}$ vector represents the zero-mean fluctuations in non-optimal directions superposed on $\overline{\underline{W}^\varepsilon}$.

The remainder of this sub-section presents a semi-empirical estimation on the norm of the $\underline{W}^{\varepsilon,\nu}$ vector; an analysis of the norm of the moving average part of $\underline{W}^\varepsilon$ is presented later on in this chapter.

The average norm of the $\underline{W}^{\varepsilon,\nu}$ vector component is by approximation given by

$$\overline{|\underline{W}^{\varepsilon,\nu}|} \approx |[\sigma_1^{w\varepsilon}, \sigma_2^{w\varepsilon}, \ldots, \sigma_{N_{in}}^{w\varepsilon}]^T|. \tag{3.12}$$

where:

$\sigma_i^{w\varepsilon}$ is the standard deviation of the ith element of $\underline{W}^{\varepsilon}$

N_{in} is the number of inputs of the neuron

To calculate the standard deviation of any element of the $\underline{W}^{\varepsilon,\nu}$ vector, $\sigma_i^{w\varepsilon}$, the probability density function of the element must be known. This probability distribution function is a function of the weight vector of the neuron and of the distribution of training vectors $\underline{U}$ in the training set:

$$\mathcal{P}(\underline{W}_i^{\varepsilon,\nu}) = \mathcal{F}\left(\mathcal{P}_u(\underline{U}), \beta^h, \beta^{bias}, \underline{W}^{\varepsilon}\right) \tag{3.13}$$

The distribution function of any element of the $\underline{W}^{\varepsilon,\nu}$ vector for subsequent adaptation on $\mathcal{N}$ training examples can now be derived by convoluting $\mathcal{N}$ single probability density functions. If the single probability density function (3.13) were independent of $\underline{W}^{\varepsilon,\nu}$, the *Central Limit Theorem* (CLT) [5] could be applied to derive an approximation for the superposition of $\mathcal{N}$ identical distribution functions. However, because of the dependency of the single distribution density function on the $\underline{W}^{\varepsilon}$ vector, see (3.13), the CLT cannot be used and a simple expression for $\sigma_i^{w\varepsilon}$ cannot be derived mathematically.

However, a *semi-empirical expression* for the standard deviation $\sigma_i^{w\varepsilon}$ can be given if both $\overline{|\underline{W}^{\varepsilon,\nu}|} \ll \beta^h$ and if the boundaries of the two classes in the training set are relatively smooth. This empirical expression is based on the CLT: the variance of any element of $\underline{W}^{\varepsilon,\nu}$ is approximated by

$$\sigma_i^{w\varepsilon} \approx \sigma_i^{w\varepsilon 0} \sqrt{N}. \tag{3.14}$$

where

$\sigma_i^{w\varepsilon 0}$ is the variance $\sigma_i^{w\varepsilon}$ under assumption of $|\underline{W}^{\varepsilon,\nu}| = 0$

N is an estimation for the number of "superposed" distributions

The first factor on the right hand side of (3.14) is

$$\sigma_i^{w\varepsilon 0} = \eta \left(\frac{1}{P} \sum_{p=1}^{P} \left(D^p - f(\underline{W}\cdot\underline{U}^p)\right)^2 f'(\underline{W}\cdot\underline{U}^p)^2 (U_i^{F,p})^2 \right)^{\frac{1}{2}}. \tag{3.15}$$

The number of "superposed" probability density function N, is empirically estimated to equal the ratio between (a) the difference between the weighted input that is required to obtain the target response and the actual weighted input and (b) the maximum adaptation of the inproduct $\underline{W}^{\varepsilon,\nu}\cdot\underline{U}^F$ with initially $\underline{W}^{\varepsilon,\nu} = \underline{0}$. This estimation is given is formula in (3.16):

$$N \approx \frac{f^{-1}(D) - \underline{W}\cdot\underline{U}\big|_{\underline{W}^{\varepsilon,\nu,0}=\underline{0}}}{sup(\Delta|\underline{W}^{\varepsilon,\nu}\cdot\underline{U}|)\big|_{\underline{W}^{\varepsilon,\nu,0}=\underline{0}}} \tag{3.16}$$

in this estimation:

$\underline{W}^{\varepsilon,\nu,0}$ denotes the $\underline{W}^{\varepsilon,\nu}$ before adaptation on the next training example

the $sup(\cdot)$ operator calculates the maximum.

In words, N approximates the minimum number of adaptations that is required to "occupy" the difference between the target weighted input and the weighted input by an inproduct $\underline{W}^{\varepsilon,\nu}\cdot\underline{U}$, starting at $\underline{W}^{\varepsilon,\nu}=\underline{0}$. The equation for N can be written as

$$N \approx \frac{f^{-1}(D)-\alpha^{h}_{min}\beta^{h}-\alpha^{bias}\beta^{bias}}{\eta\, sup\left((D-f(\alpha^{h}\beta^{h}+\alpha^{bias}\beta^{bias}))\, f'(\alpha^{h}\beta^{h}+\alpha^{bias}\beta^{bias})\,|\underline{U}^{F}|^{2}\right)}. \tag{3.17}$$

An illustration

In figure 3.2, the simulated and calculated $\overline{|\underline{W}^{\varepsilon,\nu}|}$ are shown for a linearly separable training set with relatively smooth boundaries for both classes. The correspondence between simulations and calculations is typical for all tested training sets (including non-linearly separable sets).

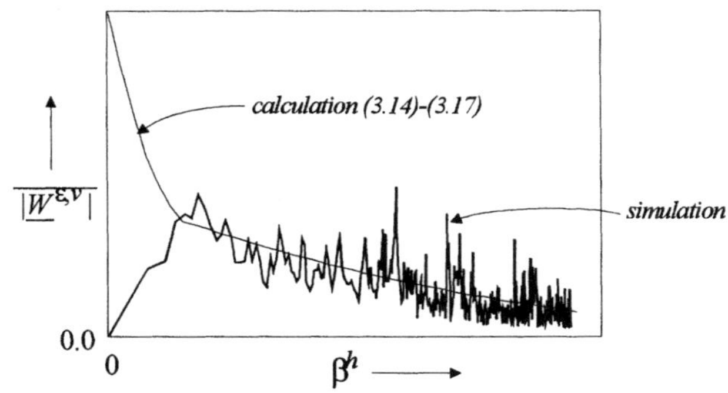

Fig. 3.2 Typical correspondence between calculated and simulated $\overline{|\underline{W}^{\varepsilon,\nu}|}$

In the beginning of training, the simulated and calculated curves in figure 3.2 are not matching, which can be explained as follows. The simulations were done with zero initial weights and zero variance for the elements of $\underline{W}^{\varepsilon,\nu}$. During training, the simulated variance increases relatively slowly because of the calculation of the relatively slow estimation of the variance of the elements of $\underline{W}^{\varepsilon}$. The mathematical expressions (3.14) to (3.17) are however based on the assumption that the "available" weighted input (the nominator of (3.17)) is completely used for $\underline{W}^{\varepsilon,\nu}\cdot\underline{U}^{F}$, which assumption is not satisfied at the beginning of training. Hence, the calculation results in a too large estimation of the variance $\overline{|\underline{W}^{\varepsilon,\nu}|}$ in the beginning of training.

3.3 The effect of the learning rate on learning

In this section, the effect of the learning rate on the learning behavior of single neurons is discussed. It is shown that for the adaptation of the weight vector in the local optimal direction, the learning rate merely acts as a scale factor. The variance of the weight vector from this local optimal direction has, in contrast to the adaptation of β^h, a square-root dependence on the learning rate. Simulation results illustrate the conclusions of this section.

Effect of η on the adaptations of β^h and β^{bias}

Assuming a negligibly small $\underline{W}^{\varepsilon}$ vector component, it follows from (3.8) to (3.10) that the adaptation of the weight vector in the locally optimal direction, $\underline{B}^h$, and the adaptation of the bias input related weight are proportional to the learning rate η. The training time to reach a certain β^h is therefore inversely proportional to this η. In figure 3.3a, the inverse of the number of training examples required to reach a certain level of β^h is plotted as a function of η for both simulation results and calculation results from (3.8) to (3.10). The training set used for the calculations and simulations is linearly separable, contains a large number of examples (approximates a continuous training example distribution) and has smooth boundaries.

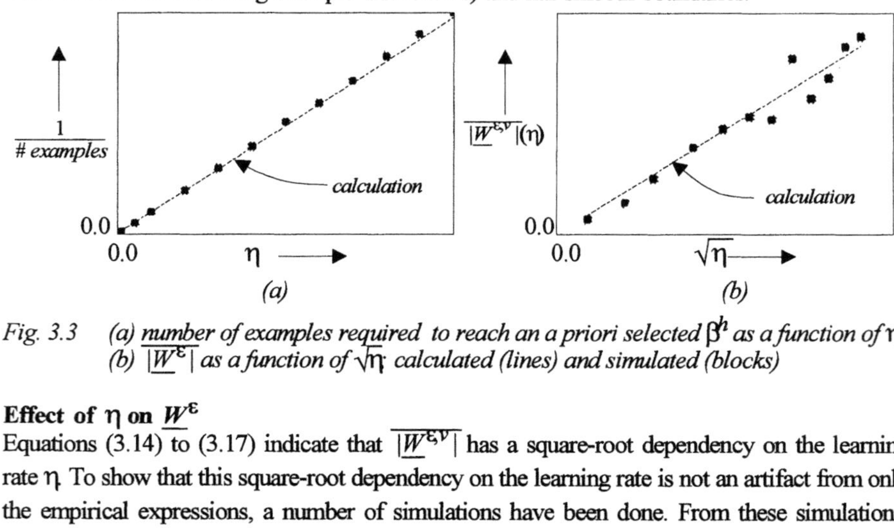

Fig. 3.3 (a) *number of examples required to reach an a priori selected* β^h *as a function of* η
(b) $\overline{|\underline{W}^{\varepsilon}|}$ *as a function of* $\sqrt{\eta}$: *calculated (lines) and simulated (blocks)*

Effect of η on $\underline{W}^{\varepsilon}$

Equations (3.14) to (3.17) indicate that $\overline{|\underline{W}^{\varepsilon,\nu}|}$ has a square-root dependency on the learning rate η. To show that this square-root dependency on the learning rate is not an artifact from only the empirical expressions, a number of simulations have been done. From these simulations, curves similar to the simulated curve in figure 3.2 resulted. After smoothening of the curves, the $\overline{|\underline{W}^{\varepsilon,\nu}|}$ corresponding to a specific a priori selected β^h were estimated. The value of the a priori selected β^h is not important in this context because it only serves as a reference point in training. Figure 3.3b shows the resulting simulated and calculated $\overline{|\underline{W}^{\varepsilon,\nu}|}$ as a function of the square root of η for the same training set as used for figure 3.3a.

In figure 3.3b, the small differences between the simulated and calculated results are mainly resulting from inaccurate estimation of the $\overline{|\underline{W}^{\varepsilon,\nu}|}$ corresponding to a specific β^h.

3.4 The effect of scaling η and $\underline{U}$

The adaptation of both β^h and β^{bias} are given in (3.8) to (3.10). It follows that under condition of equal levels of performance, the adaptation of β^h and β^{bias} scales linearly with the scale factor of the input vectors and scales linearly with the learning rate η. In formula, for equal levels of performance of the neuron:

$$\Delta\beta^h \propto \eta S_u \quad \text{and} \quad \Delta\beta^{bias} \propto \eta S_u . \tag{3.18}$$

where

S_u denotes the scale factor of the input vectors $\underline{U}$

The level of performance of the neuron is directly related to the response of the neuron on every training example. With the vector decomposition method, the level of performance is then directly related to the product $\alpha^h \beta^h$ for every example (neglecting $\underline{W}^\varepsilon$ and assuming that the actual hyperplane approximately coincides with the attractor hyperplane). It follows from (2.15) that α^h scales linearly with S_u. It can be concluded that when scaling the input signals, the β^h corresponding to a certain level of performance scales inversely proportional,

$$\beta^h_{performance} \propto \frac{1}{S_u}. \tag{3.19}$$

It follows directly from (3.18) and (3.19) that the adaptation of β^h, related to its optimal value, scales as (neglecting the effect of $\underline{W}^\varepsilon$):

$$\Delta\beta^h/\beta^h \propto \frac{1}{\eta S_u^2} \tag{3.20}$$

With (3.14) to (3.17) it follows that the $\underline{W}^{\varepsilon,\nu}$ scales linearly with the square-root of the learning rate:

$$|\underline{W}^\varepsilon| \propto \frac{1}{\sqrt{\eta}} \tag{3.21}$$

It can be concluded that under condition of a constant product $S_u^2\eta$, it follows that the variance of $\underline{W}^\varepsilon$ scales identically as $\beta^h_{performance}$. Therefore, in order to have both a constant $\Delta\beta^h/\beta^h$ and a constant $|\underline{W}^{\varepsilon,\nu}|/\beta^h$ and hence an unchanged learning behavior of the neuron, the product $S_u^2\eta$ must be kept constant. Note that an unchanged learning behavior includes that the training time to reach a specific level of performance be unchanged.

3.5 The effect of bias-input signal on learning: simple case

In feed-forward neural networks, the bias input signal of the neurons is usually taken unity. The value of this bias input however directly affects the average weight adaptation of the neuron. In this section, an analysis of the dependency of $\Delta\beta^h$ on α^{bias} is presented for a simplified case; the general case will be analyzed in section 3.6.

Assumptions

For the analysis in this section, the effect of $\underline{W}^\varepsilon$ on the weighted input is assumed to be negligibly small. This is the case if both the effect of the (moving) average part of $\underline{W}^\varepsilon$ *and* the effects of the $\underline{W}^{\varepsilon,\nu}$ on the weighted input of the neuron is negligibly small. It will be shown in section 3.6 that the moving average part of $\underline{W}^\varepsilon$ can be forced to zero during the entire training by shifting of the training set in input space over some specific and fixed $\underline{U}^F$ vector. Furthermore, it was shown in section 3.3 that the effect of the variance of the $\underline{W}^\varepsilon$ vector on the total weighted input is small for sufficiently small learning rates η.

Analysis

During training, the hyperplane of the neuron converges towards the attractor hyperplane. This attractor hyperplane provides by definition a locally optimal performance on the training set, see also section 2.3. Because the training set can be positioned at an arbitrary position in the input space, the attractor hyperplane generally does not include the origin of input space, see figure 3.4.

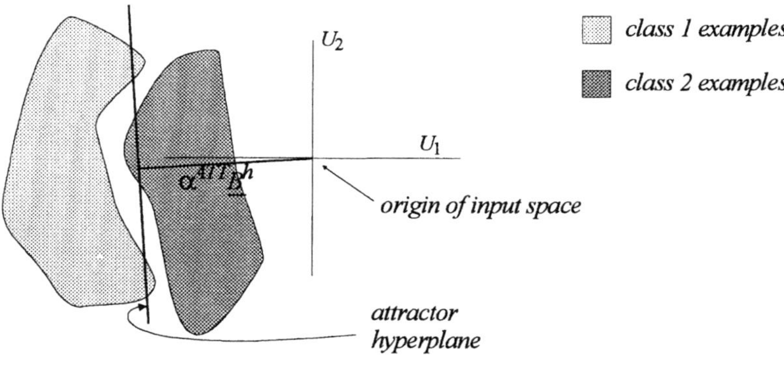

Fig. 3.4 Arbitrary training set for which the effect of $\underline{W}^\varepsilon$ is minimum (see section 3.6)

Introducing an input vector $\underline{U}^{ATT} = \alpha^{ATT}\underline{B}^h + \alpha^{bias}\underline{B}^{bias}$ which lies in the attractor hyperplane, it follows from (2.4) and (2.17) that (under condition of negligibly small $\underline{W}^\varepsilon$)

$$\alpha^{ATT}\beta^h \approx -\alpha^{bias}\beta^{bias}. \tag{3.22}$$

For the analysis of the effect of the bias signal applied to the neuron on the learning behavior, we now propose to decompose α^h as

$$\alpha^h = \alpha^{h,v} + \alpha^{ATT}. \tag{3.23}$$

In a loose way, the α^{ATT} is the average part of the α^h corresponding to training examples that result in weight adaptation; the $\alpha^{h,v}$ is then the zero-mean difference between α^h and α^{ATT}. After substitution of (3.23) in (3.8), the adaptation of β^h can be rewritten into

$$\Delta\beta^h \approx \Delta\beta^{bias}\frac{\alpha^{ATT}}{\alpha^{bias}} + \frac{\eta}{P}\sum_{p=1}^{P}\left(D^p - f(\underline{W}\cdot\underline{U}^p)\right) f'(\underline{W}\cdot\underline{U}^p)\,\alpha^{h,v,p}. \tag{3.24}$$

Because the actual weight vector attracts towards the weight vector attractor, it follows from (3.22) that, under condition of a negligibly small $\underline{W}^{\varepsilon}$,

$$\frac{\Delta\beta^h}{\Delta\beta^{bias}} \approx -\frac{\alpha^{bias}}{\alpha^{ATT}}. \tag{3.25}$$

In words, (3.25) means that the actual hyperplane is (and stays) on the average close to the attractor hyperplane. Substituting (3.25) in (3.24) yields

$$\Delta\beta^h \approx \frac{\eta'}{P}\sum_{p=1}^{P}\left(D^p - f(\underline{W}\cdot\underline{U}^p)\right) f'(\underline{W}\cdot\underline{U}^p)\,\alpha^{h,v,p} \tag{3.26a}$$

with

$$\eta' = \eta\left(\frac{\alpha^{bias^2}}{\alpha^{ATT^2} + \alpha^{bias^2}}\right). \tag{3.26b}$$

Equations (3.26a) and (3.26b) indicate that, for any training set, the adaptation of the weight vector in the local optimal direction depends on the distance between the attractor hyperplane and the origin of input space *and* depends on the magnitude of the bias input of the neuron. An illustration of this phenomenon is presented in the following sub-section.

An illustration

Figure 3.5 shows β^h as a function of the number of presented training examples for three different ratios between α^{bias} and α^{ATT}. The figure shows for the three ratios both simulation results using a neural network simulation program and shows results from (3.26).

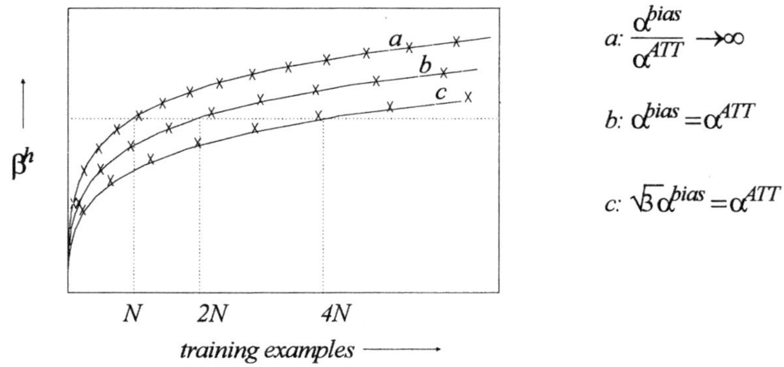

Fig. 3.5 β^h *as a function of the number of training examples for three values of* α^{ATT}*; simulated (x's) and calculated (lines)*

According to (3.26b), curves *a*, *b* and *c* require respectively minimum training time, twice the minimum and four times the minimum training time to reach an a priori selected level of β^h. This relation is clearly illustrated by the figure.

The training set used for the simulations and calculations is linearly separable and has smooth boundaries. However, as no information on the separability of the training set or on the smoothness of the set is used in the derivation of (3.26), the equation also hold for non-separable sets. The only restriction on the training set is that the effect of $\underline{W}^{\varepsilon}$ on the weighted input of the neuron be small. The next section of this chapter presents an analysis which is similar to the one in this section, but for a training set for which the effect of $\underline{W}^{\varepsilon}$ is not negligibly small.

3.6 The effect of bias-input signal on learning: general case

In the previous section of this chapter, an analysis of the effect of the bias input signal on weight adaptation of neurons was presented. The analysis in section 3.5 is restricted to the learning behavior of neurons on training sets for which the effect of $\underline{W}^{\varepsilon}$ on the learning behavior is negligibly small.

In this section, an analysis of the dependency of $\Delta\beta^h$ on α^{bias} is presented for a general case; no restrictions on the training set are required. The only assumption to be made in this section is that the effect of $\underline{W}^{\varepsilon,v}$ on the weighted input is small. For normal learning rates, this condition is satisfied. For large learning rates η, the effect of the $\underline{W}^{\varepsilon,v}$ vector component should be taken into account. This, however, is beyond the scope of this chapter.

Analysis

As shown in section 3.5, the weight associated with the bias input of the neuron adapts to a non-zero value if $\alpha^{ATT} \neq 0$, i.e. if the attractor hyperplane does not contain the origin of the input space. For training sets for which the attractor hyperplane does not contain the origin of input space, there may be a non-zero average $\underline{W}^{\varepsilon}$ weight vector component, which can be explained as follows. Due to the adaptation of the bias related weight $\beta^{bias}\underline{B}^{bias}$ on the *constant* bias input signal $\alpha^{bias}\underline{B}^{bias}$, the $\underline{W}^{\varepsilon}$ vector component tends to adapt in the direction of the *constant* part of the $\underline{U}^F$ vector component.

In the analysis in this section, it appears to be useful to decompose the $\underline{U}^F$ vector into two vectors:

$$\underline{U}^F = \underline{U}^{F,v} + \underline{U}^{F,const}. \tag{3.27}$$

where

$\underline{U}^{F,const}$ represents the constant (average) part of $\underline{U}^F$ (see below)

$\underline{U}^{F,v}$ represents the difference between $\underline{U}^F$ and $\underline{U}^{F,const}$

The constant part of the $\underline{U}^F$ vector is chosen in such a way that $\underline{U}^{F,v}$ satisfies

$$\frac{\eta}{P}\sum_{p=1}^{P}\left(D^p - f(\alpha^{h,p}\beta^h + \alpha^{bias}\beta^{bias})\right)f'(\alpha^{h,p}\beta^h + \alpha^{bias}\beta^{bias})\,\underline{U}^{F,v,p} = \underline{0}. \tag{3.28}$$

In formula, the $\underline{U}^{F,const}$ vector satisfies:

$$\left\{\exists\underline{U}^{F,const} \in \mathbb{R}^N \,\middle|\, \frac{\eta}{P}\sum_{p=1}^{P}\left(D^p - f(\alpha^{h,p}\beta^h + \alpha^{bias}\beta^{bias})\right)f'(\alpha^{h,p}\beta^h + \alpha^{bias}\beta^{bias})\left(\underline{U}^{F,p} - \underline{U}^{F,const}\right) = \underline{0}\right\} \tag{3.29}$$

A general training set with non-zero norms for the constant part of all three components of the average input vector is shown in figure 3.6. The two vector components that point towards the average point in input space are given in (3.23) and (3.27).

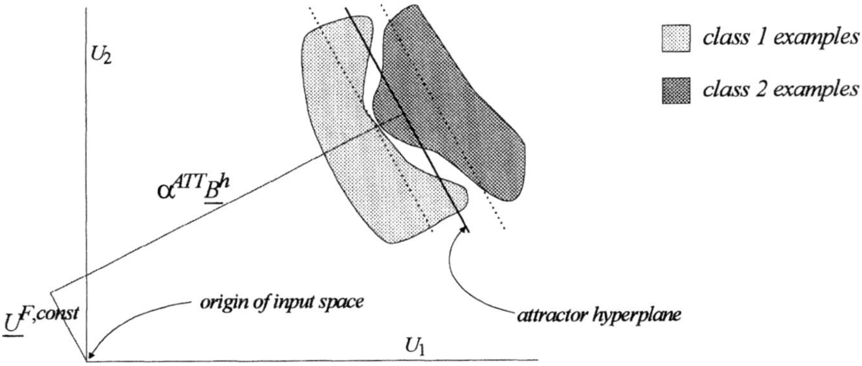

Fig. 3.6 Arbitrary (non-linearly separable) training set and decomposition of the vector that points towards the average of the training vectors

As mentioned earlier in this section, the adaptation of β^{bias} on α^{bias} results in adaptation of $\underline{W}^{\varepsilon}$ on $\underline{U}^{F,const}$. With (3.11) and (3.27), it follows that

$$\left(\Delta\overline{\underline{W}^{\varepsilon}}\right)_{ufconst} = \underline{U}^{F,const} \cdot \frac{\Delta\beta^{bias}}{\alpha^{bias}} . \tag{3.30}$$

With the definition of $\underline{U}^{F,const}$ in (3.29), it follows that the average adaptation of the $\underline{W}^{\varepsilon}$ vector due to only the $\underline{U}^{F,v}$ vector components is zero for any state of the neuron in which $\underline{W}^{\varepsilon}=\underline{0}$. It is straight forward to show that for $\underline{W}^{\varepsilon}\neq\underline{0}$ (under assumption of double threshold pattern classification tasks), the adaptation on examples that tend to reduce the norm of $\underline{W}^{\varepsilon}$ is enlarged while the adaptation on examples that tend to increase $|\underline{W}^{\varepsilon}|$ is decreased. This means that the average adaptation of the $\underline{W}^{\varepsilon}$ vector component due to only $\underline{U}^{F,v}$ is such that it tends to reduce the norm $|\underline{W}^{\varepsilon}|$:

$$\left(\Delta\overline{\underline{W}^{\varepsilon}}\right)_{w\varepsilon} = \frac{\eta}{P}\sum_{p=1}^{P}\left(D^p - f(\underline{W}\cdot\underline{U}^p)\right) f'(\underline{W}\cdot\underline{U}^p)\,\underline{U}^{F,p} . \tag{3.31}$$

The total adaptation of $\overline{\underline{W}^{\varepsilon}}$ is the sum of the $\underline{U}^{F,const}$ induced and the $\overline{\underline{W}^{\varepsilon}}$ caused adaptation. The first tends to increase the norm of the $\overline{\underline{W}^{\varepsilon}}$ vector component and the second tends to reduce its norm,

$$\Delta\overline{\underline{W}^{\varepsilon}} = \left(\overline{\underline{W}^{\varepsilon}}\right)_{ufconst} + \left(\overline{\underline{W}^{\varepsilon}}\right)_{w\varepsilon} . \tag{3.32}$$

with

$$\frac{d\left|\left(\overline{\underline{W}^{\varepsilon}}\right)_{ufconst}\right|}{dt} \geq 0 \quad \text{and} \quad \frac{d\left|\left(\overline{\underline{W}^{\varepsilon}}\right)_{w\varepsilon}\right|}{dt} \leq 0$$

By definition, the attractor hyperplane of a neuron represents the hyperplane that corresponds to an optimal performance for the state of the neuron. This means that with learning rules that minimize a cost function the actual hyperplane attracts towards the attractor hyperplane. However, due to the non-zero norm of the average $\overline{\underline{W}^{\varepsilon}}$ vector, the actual hyperplane will be rotated with respect to the attractor hyperplane.
For the rotated hyperplane, the minimum in the energy is obtained when approximately including the input vector

$$\underline{U}^{ATT} = \alpha^{ATT}\underline{B}^{h} + \alpha^{bias}\underline{B}^{bias} + \underline{U}^{F,const} \tag{3.33}$$

in the actual hyperplane[1]. Because of this, the weighted input of the neuron on the $\underline{U}^{ATT}$ vector is by approximation zero:

$$\alpha^{ATT}\beta^{h} + \alpha^{bias}\beta^{bias} + \underline{U}^{F,const}\cdot\overline{\underline{W}^{\varepsilon}} \approx 0\,. \tag{3.34}$$

Again, after adaptation (3.34) will still be satisfied when assuming that the attractor hyperplane is (at least quasi-) stationary. With (3.30) it can be derived that the adaptation of the bias-input-related weight is given by

$$\Delta\beta^{bias} \approx -\alpha^{bias}\,\frac{\alpha^{ATT}\Delta\beta^{h} + \underline{U}^{F,const}\cdot\left(\Delta\overline{\underline{W}^{\varepsilon}}\right)_{W\varepsilon}}{\alpha^{bias^2} + |\,\underline{U}^{F,const}\,|^2}\,. \tag{3.35}$$

Substitution of (3.35) in (3.24) yields now an equation that describes the adaptation of the "correct knowledge" part of the weight vector:

$$\Delta\beta^{h} \approx \frac{\eta'}{P}\sum_{p=1}^{P}\left(D^{p} - f(\underline{W}\cdot\underline{U}^{p})\right)f'(\underline{W}\cdot\underline{U}^{p})\,\alpha^{h,\nu,p} - \frac{\alpha^{ATT}\,\underline{U}^{F,const}\cdot\left(\Delta\overline{\underline{W}^{\varepsilon}}\right)_{W\varepsilon}}{\alpha^{bias^2} + |\,\underline{U}^{F,const}\,|^2 + \alpha^{ATT^2}}\,, \tag{3.36}$$

$$\text{with } \eta' = \eta\left(\frac{\alpha^{bias^2} + |\underline{U}^{F,const}|^2}{\alpha^{bias^2} + |\,\underline{U}^{F,const}\,|^2 + \alpha^{ATT^2}}\right).$$

[1]Note that by definition of the various components in (3.33), the vector $\underline{U}^{ATT}$ lies in the attractor hyperplane of the neuron.

Direct conclusions
It can be concluded from (3.30) and (3.36) that there exist two cases for which $\overline{\underline{W}^{\varepsilon}}$ *either* equals and stays zero *or* strictly monotonuously decreases to zero during training (depending on the initial $\underline{W}^{\varepsilon}$).

- Firstly, in the case that the attractor hyperplane contains the origin of input space, or in formula $\alpha^{ATT}=0$, both the β^{bias} and $\Delta\beta^{bias}$ decrease to zero. As a result of this, the leftmost term on the right hand side of (3.32) decreases to zero according to (3.30).
- Secondly, $\overline{\underline{W}^{\varepsilon}}$ remains zero during training if all elements of $\underline{U}^{F,const}$ are zero.

In all other cases, the weight vector will build up a non-zero norm for the $\overline{\underline{W}^{\varepsilon}}$ vector. This means effectively that the weight vector on the average does not only adapt in the local optimal direction; the weight vector follows a curved path in weight space from the initial point towards the weight vector attractor.

An illustration
An illustration of the adaptation of the $\overline{\underline{W}^{\varepsilon}}$ vector as a function of the β^{h} is presented in figure 3.7. For a linearly separable training set with smooth boundaries, the dynamics of the weight vector of a neuron have both been calculated using (3.32) and (3.36) and have been simulated with a neural network simulation program. The correspondence between the simulation and the calculation in figure 3.7 is typical.

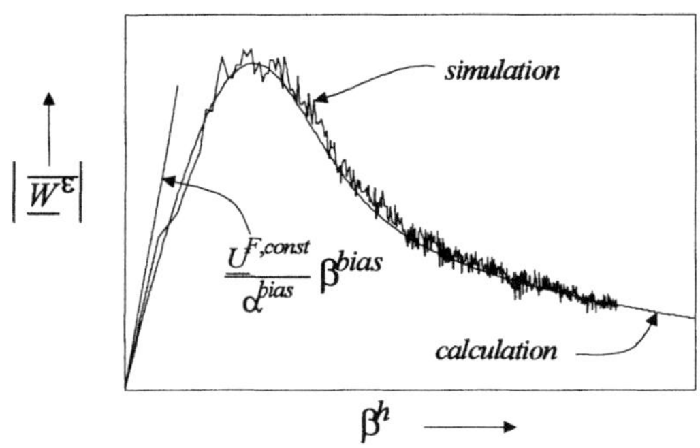

Fig. 3.7 Simulated and calculated norm of $\overline{\underline{W}^{\varepsilon}}$ as a function of β^{h}

In the beginning of training, $\overline{\underline{W}^{\varepsilon}}$ adapts mainly only according to (3.30) which is illustrated by the line $\beta^{bias}\underline{U}^{F,const}/\alpha^{bias}$ in figure 3.7. After initial training, (3.31) becomes more and more dominant over (3.30), resulting in an eventual decrease of the norm of $\overline{\underline{W}^{\varepsilon}}$ during training.

3.7 Conclusions

The Vector Decomposition Method which was introduced in chapter 2 has been used analyze the adaptation of weight vector of neurons in single-layer feed-forward neural networks. Easy-to-read equations are presented that describe the effect of the bias input signal of the neuron, the learning rate η, and scaling input vectors on the adaptation of the weight vector. It is shown that in general, the weight vector follows during training a curved path in weight space from its initial point towards its eventual point. Application of the equations that are derived in this chapter with respect to the effect on training time of a few hardware-implementation-related problems are discussed in the next four chapters for single-layer feed-forward neural networks. After these chapters, the Vector Decomposition Method is used to analyze the learning behavior of multi-layer feed-forward neural networks.

3.8 References

[1] T.M.Heskes and B.Kappen, "Learning processes in neural networks", *Physical Review A*, vol. 44, 2718-2726, 1991

[2] T.M.Heskes, E.T.P.Slijpen and B.Kappen, "Learning in neural networks with local minima", *Physical Review A*, vol. 46, 5221-523, 1992

[3] R.P.Lippmann, "An Introduction into Computing with Neural Nets", *IEEE ASSP Mag.*, vol. 4, 4-22, 1987

[4] M.L.Minsky and S.A.Papert, *Perceptrons, An Introduction to Computational Geometry*, Expanded Edition, Cambridge, MA.: MIT Press, 1988

[5] A.Papoulis, *Probability, Random Variables and Stochastic Processes*, New York: McGraw-Hill, 1984

[6] D.B.Parker, Learning Logic, Technical Report 47, MIT, 1985

[7] D.E.Rumelhart, G.E.Hinton and R.J.Williams, "Learning internal representations by error propagation", in *Parallel Distributed Processing*: Exploration of the microstructure of cognition, vol. 1, Cambridge, MA.: MIT Press, 1986

[8] E.D.Sontag and H.J.Sussmann, "Back Propagation Separates Where Perceptrons Do", *Neural Networks*, vol.4, 243-249, 1991

[9] J.J.Shynk and N.J.Bershad, "Steady-State Analysis of a Single-Layer Perceptron Based on a System Identification Model with Bias Terms", *IEEE Trans. Neural Networks*, vol.38, 1030-1042, 1991

[10] J.J.Shynk and N.J.Bershad, "Stationary Points and Performance Surfaces of a Perceptron Learning Algorithm for a Nonseparable Data Model", In *Proceedings of the IJCNN* (Baltimore), vol. 2, 133-139, 1992

[11] P.J.Werbos, "Backpropagation: Past and Future", In *Proceeding of the IJCNN* (San Diego), CA., vol. 1, 343-353, 1988

Chapter 4

Unipolar Input Signals in Single-Layer Feed-Forward Neural Networks

4.1 Introduction

In neural networks, signal processing is in principle performed by simple processors (neurons) operating in parallel. Implementing neural networks in analog hardware seems therefore natural. In the aim at reducing the chip size and the power consumption, which requires simplification of the signal processing blocks, the convergence of the neural network generally slows down. In analog hardware implementations of neural networks, two quadrant multipliers have several advantages over four quadrant multipliers:

- two-quadrant multipliers (including their drivers) generally have a lower circuit complexity
- two-quadrant multipliers generally have a lower power consumption than equally fast four quadrant multipliers

Furthermore, in case of using two-quadrant multipliers for the synapses, the multiplier in the weight-adaptation circuit can also be a two-quadrant multiplier. However, to be able to use two-quadrant multipliers, the input signals of the neural network need to be unipolar. In this chapter, the VDM will be used to analyze the effect of the consequences on training time of a translation of the training set in input space, such that unipolar input signals for the neural network are obtained. Conditions guaranteeing minimum training time are presented as well.

4.2 Translations towards unipolar input signals

In analog hardware feed-forward neural networks, the input signals can be either bipolar or unipolar. In the first case, the synapses and the multipliers for the weight adaptation are represented by four-quadrant multipliers [1-3],[5-10],[13],[15],[17],[18],[20-22] whereas in the second case both the synapses and the weight-update multipliers can be implemented using two-quadrant multipliers [4],[11],[14],[16],[19]. Because two-quadrant multipliers (including wiring) are generally smaller and consume less power than equally fast four-quadrant multipliers, it is efficient from an implementation point of view to shift all input signals of the neural net in input space to obtain unipolar input signals.

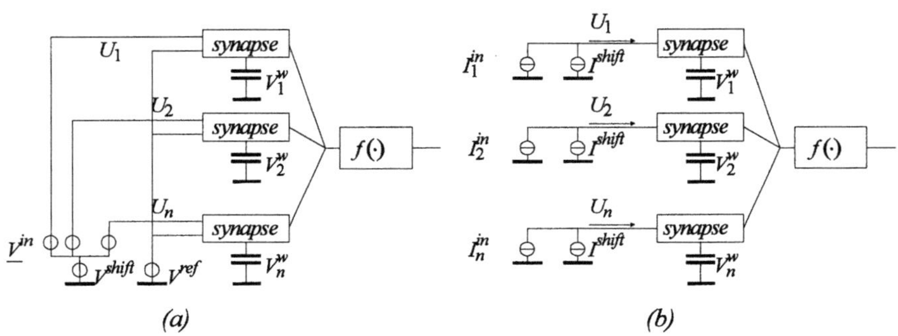

Fig. 4.1 Two analog implementations of neurons with translated input signals: (a) voltage inputs and (b) current inputs

Figure 4.1a illustrates the translation of an input vector $\underline{V}^{in}$ to the actual input vector of the neuron $\underline{U}$ for an analog hardware implementations of a neuron with voltage input; note that both V^{shift} and V^{ref} may be used to translate the training set in input space. Figure 4.1b shows the translation of the input vector $\underline{I}^{in}$ to $\underline{U}$ for a neuron with input signals in the current domain. The elements of the vector $\underline{U}$ are the actual input signals for the synapses and hence for the neuron. For both neurons in figure 4.1, the weights are stored as voltages V^{w} across capacitors. For digital hardware implemented neural nets and software neural networks, unipolar input signals correspond to input signals $U_j \in \mathbb{R}^+$ and $U_j \in \mathbb{N}$ for nets operating with floating point inputs and integer values respectively.

This chapter presents analyses of the effects of translating the training set in input space to obtain unipolar input signals. Conditions guaranteeing minimum training time for learning the translated training set follow from these analyses.

4.2.1 Centre-of-gravity

When translating the training set in input space, the adaptation of the weight vector in the local optimal direction and the deviation of the weight vector from this local optimal direction generally changes, see section 3.6. In (3.32) and (3.36), this change in adaptations as a consequence of a training set translation in input space can be expressed in terms of α^{ATT}, $\underline{U}^{F,const}$ and α^{bias}. It has been shown in section 3.6 that the attractor hyperplane includes the point

$$\underline{U}^{CG} = \alpha^{ATT}\underline{B}^{h} + \alpha^{bias}\underline{B}^{bias} + \underline{U}^{F,const}, \tag{4.1}$$

which points to a fixed spot in the training set. The point $\alpha^{ATT}\underline{B}^{h} + \underline{U}^{F,const}$ will be denoted as the *centre-of-gravity* of a training set. In a loose way and under the assumption that the attractor hyperplane is close to the actual hyperplane, the centre-of-gravity can be approximated by the point in input space that is the average of all training examples that result in weight adaptation. It follows from (3.36) that if this centre-of-gravity coincides with the origin of input space, the weight vector adapts maximally in the local optimal direction, $\underline{B}^{h}$, and the deviation form this local optimal direction is minimum.

Because the centre-of-gravity is fixed with respect to the training set, the translation of the training set corresponds to an identical translation of the centre-of-gravity. The effect of a training set translation in input space can easily be calculated by substituting the translated centre-of-gravity in (3.36).

4.2.2 Minimum training time for fixed learning rate η

After translating a training set in input space to obtain unipolar input signals, all vector components of the centre-of-gravity will be non-zero. With (3.36), this results in non-optimum adaptation of the weight vector in the optimal direction and hence a non-minimum training time. Due to the non-linear and recursive nature of (3.30) to (3.32), the change in training time required to reach a specific level of performance is heavily dependent on the training set, so no closed expression can be derived.

However, it follows from (4.1) and (2.40) that the adaptation of the weight vector in the $\underline{B}^{h}$ direction can be maximized by applying a large bias input signal to the neuron:

$$|\underline{U}^{F} + \alpha^{ATT}\underline{B}^{h}| \ll |\alpha^{bias}| \quad \Leftrightarrow \quad \eta' \approx \eta. \tag{4.2}$$

Furthermore, under condition of a large bias input signal, the average deviation of the weight vector from its local optimal direction is minimized, as follows from (3.36); the second term on the right hand side of (3.36) will therefore also be minimized. It can be concluded that to minimize training time of the neuron for any training set, the bias input signal must be large compared to the norm of any input vector excluding the bias element in that vector. With (4.1), this results in the relation

$$|\alpha^{bias}| \approx |\underline{U}^{CG}| \quad \Leftrightarrow \quad \eta' \approx \eta. \tag{4.3}$$

4.2.3 Minimum training time, including scaling of η

For fast convergence of the weight vector, the learning rate must be as large as possible. However, in order to have a proper convergence of the weight vector of a neural network, the learning rate η must be small enough to ensure that any adaptation of the weighted input is small. In literature some estimations on optimum learning rates η are published [23],[24]. In general, the *maximum* usable learning rate η is proportional to the reciprocal of the maximum of the squared norm of the input vector $\underline{U}$ in order to ensure stability during training:

$$\eta \propto \frac{1}{(N_{in}-1)\hat{U}^2 + \alpha^{bias^2}} \tag{4.4}$$

where $\hat{U}$ denotes the maximum of the absolute value of any element of $\underline{U}$ (excluding the bias input signal)

N_{in} denotes the number of elements in $\underline{U}$

With (3.36) it can be derived that

$$\eta' \propto \frac{1}{(N_{in}-1)\hat{U}^2 + \alpha^{bias^2}} \left(\frac{\alpha^{bias^2} + |\underline{U}^{F,const}|^2}{\alpha^{bias^2} + |\underline{U}^{F,const}|^2 + \alpha^{ATT^2}} \right)$$

It is now straight forward to calculate that the maximum η', as a function of α^{bias}, occurs for

$$|\alpha^{bias,opt}| = \sqrt{-\underline{U}^{F,const^2} + |\alpha^{ATT}| \sqrt{(N_{in}-1)\hat{U}^2 - |\underline{U}^{F,const}|^2}} \tag{4.5}$$

This maximum is caused by mainly increasing of the η' with α^{bias} for low α^{bias}, and the subsequent dominant decrease of η due to the increase of the norm of the maximum input vector in (4.4). Note that due to a translation of the training set in input space, the training set related $\underline{U}^{F,const}$ and α^{ATT} change.

Figure 4.2 shows the virtual learning rate η' as a function of the α^{bias}. The two vertical dotted lines in the figure correspond to the two values of $\alpha^{bias,opt}$ following from (4.5).

Note however that although the virtual learning rate η' is non-zero for $\alpha^{bias}=0$ (due to a non-zero norm for the $\underline{U}^{F,const}$ component), the neuron will only learn a training set properly if the distance between the attractor hyperplane and the origin of input space is very small.

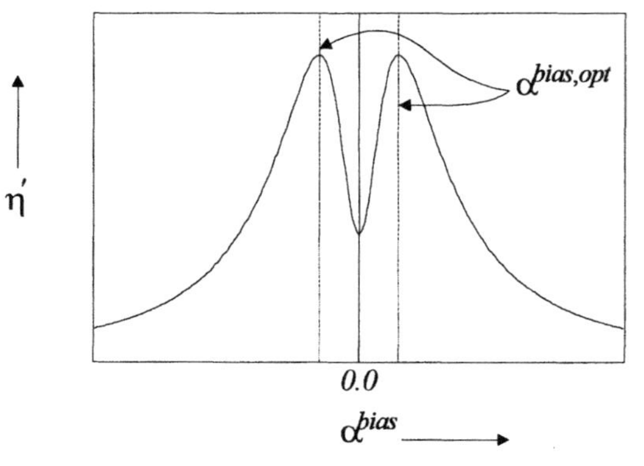

Fig. 4.2 The virtual learning rate η' as a function of α^{bias}

4.2.4 Discussion

When shifting the training set in order to allow the use of two-quadrant multipliers instead of four-quadrant multipliers, the training time generally increases. In case that the learning rate η is not changed, the training time required to reach a certain level of performance remains approximately minimum if the magnitude of the bias input signal is large compared to the norm of any training vector. If the training time is to be minimized by both changing the learning rate and by changing the bias input signal, there is an optimum value for the bias input signal of the neuron, given by (4.5). Another way to (virtually) change the bias input of the neuron is to modify the learning rate of *only* the bias-signal related weight (see section 3.4). Note that for an identical effect on training, the change of the learning rate should be the square of the factor at which the bias input signal would have been changed.

4.3 References

[1] S.Bibyk and M.Ismail, "Issues in Analog VLSI and MOS Techniques for Neural Computing", In (C.Mead and M.Ismail, eds.) *Analog VLSI Implementation of Neural Systems*, Kluwer, 103-133, 1989

[2] K.Bult and H.Wallinga, "A CMOS four-quadrant analog multiplier", *IEEE Journal of Solid-State Circuits*, vol. 21, 430-435, 1986

[3] H.A.Castro, S.M.Tam and M.A.Holler, "Implementation and Performance of an Analog Nonvolatile Neural Network", *Analog Integrated Circuits and Signal Processing*, vol. 4, 97-113, 1993

[4] D.D.Caviglia, M.Valle and G.M.Bisio, "Effects of Weight Discretization on the Back Propagation Learning Method: Algorithm Design and Hardware realization", In *Proc. of the International Joint Conference on Neural Networks* (San Diego), vol.2, 631-637, 1990

[5] M.R.Choi, *Implementation of Feedforward Artificial Neural Networks With Learning Using Standard CMOS Technology*, Ph.D. Thesis, Michigan State University, 1991

[6] J.Choi, S.H.Bang and B.J.Sheu, "A programmable Analog VLSI Neural Network Processor for Communication Receivers", *IEEE Transactions on Neural Networks*, 484-495, 1993

[7] W.C.Fang, B.J.Sheu, O.T.C.Chen and J.Choi, "A VLSI Neural Processor for Image Data Compression Using Self-Organization Networks", *IEEE tr. Neural Networks*, vol. 3, 1992

[8] F.J.Kub, K.K.Moon, I.A.Mack and F.M.Long, "Programmable Analog Vector-Matrix Multipliers", *IEEE Journal of Solid-State Circuits*, 207-214, 1990

[9] J.A.Lansner and T.Lehmann, "A Neuron- and Synapse Chip for Artificial Neural Networks", In *Proceeding of the European Solid-State Circuits Conference* (Copenhagen), 213-216, 1992

[10] J.A.Lansner and T.Lehmann, "An Analog CMOS Chip Set for Neural Networks with Arbitrary Topologies", *IEEE Transactions on Neural Networks*, 441-445, 1993

[11] J.C.Lee, B.J.Sheu, W.C.Fang and R.Chelappa, "VLSI Neuroprocessors for Video Motion Detection", *IEEE Transactions on Neural Networks*, 178-190, 1993

[12] B.Linares-Barranco, E.Sánches-Sinencio, A.Rodríguez-Vázquez and J.L.Huertas, "VLSI Implementation of a Transconductance mode Continuous BAM with On Chip Learning and Dynamic Analog Memory", In *Proceeding of the IEEE International Symposium on Circuits and Systems* (Singapore),1283-1286, 1991

[13] B.Linares-Barranco, E.Sánches-Sinencio, A.Rodríguez-Vázquez and J.L.Huertas, "A CMOS Analog Adaptive BAM with On-Chip Learning and Weight Refreshing", *IEEE Transactions on Neural Networks*, 445-455, 1993

[14] J.B.Lont and W.Guggenbühl, "Analog CMOS Implementation of a Multilayer Perceptron with Nonlinear Synapses", *IEEE Transactions on Neural Networks*, 457-465, 1992

[15] T.C.Lu, M.L.Chiang and J.B.Kuo, "A One-Transistor Synapse Circuit with an Analog LMS Adaptive Feedback for Neural Network VLSI", In *Proceedings of the IEEE International Symposium on Circuits and Systems* (Singapore), 1303-1306, 1991

[16] F.J.Pelayo, B.Pino, A.Prieto, J.Ortega and F.J.Fernandez, "Analog Implementation of a Synaptic Matrix with Alterable Weight", In *Proceedings of the 2nd International Conference on Microelectronics for Neural Networks*, Munich: Kyrill & Method Verlag, 261-267, 1991

[17] F.M.A.Salam, M.R.Choi and Y.Wang, "Artificial Neural Nets in MOS Silicon", In (Sethi & Jain, eds.) *Artificial Neural Networks and Statistical Pattern Recognition*, 243-270, 1991

[18] S.Satyanarayana, Y.P.Tsividis and H.P.Graf, "A Reconfigurable VLSI Neural Network", *IEEE Journal of Solid-State Circuits*, 67-81, 1992

[19] B.J.Sheu, "VLSI Neurocomputing with Analog Programmable Chips and Digital Systolic Array Chips", In *Proceedings of the IEEE International Symposium on Circuits and Systems* (Singapore), 1267-1270, 1991

[20] L.Song, M.I.Elmasry and A.Vanelli, "Analog Neural Network Building Blocks Based on Current Mode Subthreshold Operation", In *Proceedings of the IEEE International Symposium on Circuits and Systems* (Chicago), 2462-2465, 1993

[21] Y.Wang, "A Modular Analog CMOS LSI for Feedforward Neural Networks with On-Chip BEP Learning", in *Proceeding of the IEEE International Symposium on Circuits and Systems* (Chicago), 2744 2747, 1993

[22] R.J.Wiegerink, "A CMOS analog four-quadrant multiplier", In *Proceedings of the IEEE International Symposium on Circuits and Systems* (Singapore), 2244-2247, 1991

[23] H.C.Eaton and T.L.Oliver, "Learning Coefficient Dependence on Training Set Size", Neural Networks, vol. 5, pp. 283-288, 1992

[24] S.Roy, "Factors influencing the choice of a learning rate for a backpropagation neural network", in Proc. IEEE ICNN'94, pp. 503-507, Orlando, Fla, 1994

Chapter 5

Cross-talk in Single-Layer Feed-Forward Neural Networks

5.1 Introduction

In integrated analog neural networks the input signals are inherently subjected to mutual cross-talk via capacitances and inductances between the relatively long and close interconnection lines. Resistive coupling may occur via power supply lines, and an additive type of coupling may occur in multiplex systems used to reduce to number of I/O pins. This chapter presents an analysis of the effects of coupling between input signals on the learning behavior of single layer feed-forward neural networks. As a result of this analysis, specifications for the interconnections in a neural network can be derived.

5.2 Coupling between input signals

For feed-forward neural nets realized in analog hardware, the input signals are usually applied to the neurons through a number of interconnection lines in parallel. Because these input lines are relatively long and close to each other, capacitive and inductive coupling will occur between input signals [4],[5]. Resistive coupling may occur due to coupling via power supply lines (including ground connections which are usually applied via the resistive substrate) [3]. Furthermore, an additive type of coupling may result from multiplexing the input signals to reduce the number of input-pins [1],[2]. In this section, the effect of coupling between input signals (or cross-talk) is analyzed.

5.2.1 Analysis of the effect of coupling

It will be assumed that the coupling between input signals is linear. This means that coupling can be represented by a coupling matrix containing the coupling factors between any input signals. This matrix will be denoted as $\underline{\underline{CM}}$ and the transformed input vectors will be denoted as $\underline{U}^*$. The transformed input signal of the neuron is then

$$\underline{U}^* = \underline{\underline{CM}}\,\underline{U} \tag{5.1}$$

The model of the coupling is shown in figure 5.1.

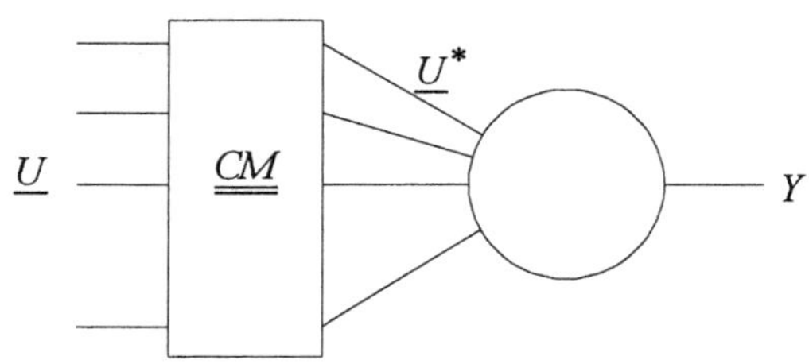

Fig. 5.1 Model of input-signal coupling: a coupling matrix

The position of the attractor hyperplane of the transformed training set is generally different from that of the original training set. As a result, the weight vector of the transformed training set is different from the weight vector that corresponds to the original set. For the weight vector corresponding to the transformed set and for its vector components we will use an asterix superscript, conform the notation in (5.1).

An illustartion of the effect of slanting a training set in input space, typically caused by coupling between input signals, is shown in figure 5.2. Figure 5.2a presents the original (unslanted) training set; for illustration purposes, the borders of the two classes are piece wise linear; most of them being either perpendicular or in parallel to the attractor hyperplane. The vector perpendicular to the attractor hyperplane in figure 5.2a is the $\underline{B}^h$ unit vector.

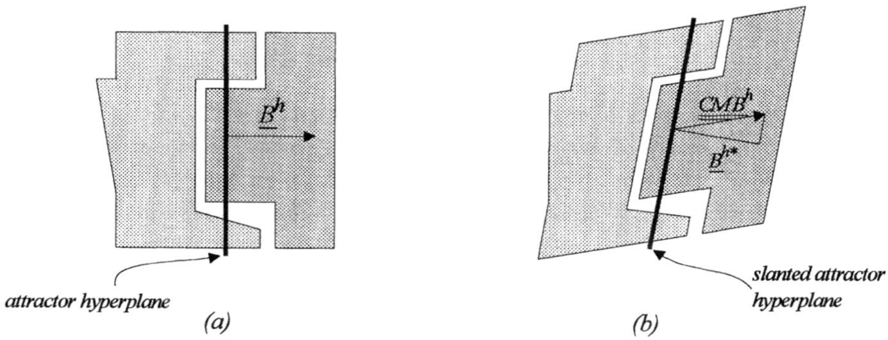

Fig. 5.2 (a) the original training set
(b) the slanted set, including $\underline{\underline{CM}}\underline{B}^h$ *and* $\underline{B}^{h*}$ *which vectors are generally not identical*

Due to coupling between input signals of the neuron, the training set (as "seen" by the neuron) is slanted in input space. For a certain coupling between input signals, this phenomenon is shown in figure 5.2b. It is clear from this figure that the slanted $\underline{B}^h$ vector, $\underline{\underline{CM}}\underline{B}^h$, is generally not perpendicular to the slanted attractor hyperplane. In the next sub-section, among others the direction of the unit vector that *is* perpendicular to the attractor hyperplane, $\underline{B}^{h*}$, is derived.

Derivation of $\underline{B}^{h*}$ and α^{h*}

It has been illustrated and argumented that the slanted $\underline{B}^h$ unit vector is generally *not* perpendicular to the slanted attractor hyperplane. In formula, this means that generally

$$\underline{B}^{h*} \neq \underline{\underline{CM}}\,\underline{B}^h.$$

Noting that by definition the attractor hyperplane of the neuron is in parallel to any $\underline{U}^F$ vector component, the direction of the $\underline{B}^{h*}$ vector component follows from the following equality:

$$\left(\underline{\underline{CM}}\,\underline{U}^F\right)\cdot\underline{B}^{h*} = 0 \; \left(= \underline{U}^{F^T}\underline{\underline{CM}}^T\underline{B}^{h*}\right).$$

It follows that the unit vector $\underline{B}^{h*}$ is

$$\underline{B}^{h*} = \frac{\underline{\underline{CM}}^{-1^T}\underline{B}^h}{\left|\underline{\underline{CM}}^{-1^T}\underline{B}^h\right|}. \tag{5.2}$$

in which equation the denominator serves to normalize the vector. By substitution of (5.2) in the equality $\alpha^{h*} = (\alpha^h \underline{\underline{CM}}\,\underline{B}^h)\cdot\underline{B}^{h*}$, it follows that the relation between the "relevant information" in the input examples (see section 2.4.2) is

$$\alpha^{h*} = \frac{\alpha^h}{|\underline{\underline{CM}}^{-1^T}\,\underline{B}^h|}\,. \tag{5.3}$$

Derivation of $\underline{U}^{F*}$

Each $\underline{U}^F$ component vector of training vectors are transformed by the coupling matrix, and is in parallel to the (now slanted) attractor hyperplane. The transformed $\underline{U}^F$ vectors will be denoted as $\underline{U}^{FF*}{=}\underline{\underline{CM}}\,\underline{U}^F$. Furthermore, because the $\underline{B}^{h*}$ unit vector is generally not perpendicular to the attractor hyperplane, an α^h-dependent $\underline{U}^{F*}$ results from the transformation of the input space. This component is the part of $\underline{\underline{CM}}\,\underline{B}^h$ that is in parallel to the attractor hyperplane, see also figure 5.2b.

The total $\underline{U}^{F*}$ vector component is

$$\underline{U}^{F*} = \underline{\underline{CM}}\cdot\underline{U}^F + \alpha^{h*}\left[|\underline{\underline{CM}}^{-1^T}\,\underline{B}^h|\,\underline{\underline{CM}}\,\underline{B}^h - \underline{B}^{h*}\right] = \underline{U}^{FF*} + \alpha^{h*}\,\underline{U}^{F\alpha*} \tag{5.4}$$

Analysis of the learning behavior

By definition, the actual hyperplane of the neuron attracts towards the attractor hyperplane during training. It has been shown in chapter 3 that under condition of a zero $\overline{\underline{W}^{\varepsilon}}$ vector component, the actual hyperplane and the attractor hyperplane will approximately coincide during training. However, the weight vector generally will have a non-zero average deviation from the local optimal direction, see (3.36). It has been shown in chapter 3 that for this general situation, the so-called centre-of-gravity is by approximation contained in the hyperplane,

$$\alpha^{ATT*}\,\Delta\beta^{h*} + \alpha^{bias*}\,\Delta\beta^{bias*} + \Delta\overline{\underline{W}^{\varepsilon*}}\cdot\underline{U}^{F,constant*} \approx 0. \tag{5.5}$$

For single neurons with *zero* coupling between the input signals, the adaptation of $\overline{\underline{W}^{\varepsilon}}$ consists of *two* components, (3.30) and (3.31). In case of *non-zero* coupling between input signals, the adaptation of the $\overline{\underline{W}^{\varepsilon*}}$ vector component is the sum of *three* components. The first term is analogous to (3.30) and increases the norm of $\overline{\underline{W}^{\varepsilon*}}$:

$$\left(\Delta\overline{\underline{W}^{\varepsilon*}}\right)_{\underline{U}^{F,constant*}} = \underline{U}^{F,constant*}\cdot\frac{\Delta\beta^{bias*}}{\alpha^{bias*}}\,. \tag{5.6}$$

Secondly, a term corresponding to (3.31) results in a decline of the norm of $\overline{\underline{W}^{\varepsilon*}}$ as has been explained in section 3.6.

$$\left(\Delta\overline{\underline{W}^{\varepsilon*}}\right)_{\underline{W}^{\varepsilon*}} = \frac{\eta}{P}\sum_p\left(D^p - f(\underline{W}^*\cdot\underline{U}^{*p})\right)f'(\underline{W}^*\cdot\underline{U}^{*p})\,\underline{U}^{F*,p}\,. \tag{5.7}$$

The third term results from the α^h-dependent $\underline{U}^{F*}$ vector component, see (5.4), and is specific for neurons with coupled input signals. This term is:

$$\left(\underline{\Delta W}^{\varepsilon*}\right)_{\alpha^{h*}} = \Delta\beta^{h*}\underline{U}^{F\alpha*}. \tag{5.8}$$

It follows with (5.5) to (5.8) that the adaptation of the bias-input related weight is

$$\Delta\beta^{bias*} \approx \frac{-\alpha^{bias*}\left[(\alpha^{ATT*}+\underline{U}^{F\alpha*}.\underline{U}^{F,constant*})\Delta\beta^{h*}+\underline{U}^{F,constant*}.\Delta\left(\underline{W}^{\varepsilon*}\right)_{W\varepsilon}\right]}{\alpha^{bias*2}+|\underline{U}^{F,constant*}|^2}. \tag{5.9}$$

Keeping in mind that the actual hyperplane attracts to the attractor hyperplane and that the attractor hyperplane is directly related to the training set (independent of the position in input space of the training set) it was derived in chapter 3 that when decomposing α^{h*} as

$$\alpha^{h*} = \alpha^{\nu*} + \alpha^{ATT*} \tag{5.10}$$

the adaptation of the norm of the weight vector in the locally optimal direction can be rewritten into

$$\Delta\beta^{h*} \approx \Delta\beta^{bias*}\frac{\alpha^{ATT*}}{\alpha^{bias*}} + \frac{\eta}{p}\sum_{p}\left(D^p - f(\underline{W}^*\cdot\underline{U}^{*p})\right)f'(\underline{W}^*\cdot\underline{U}^{*p})\,\alpha^{\nu*,p}. \tag{5.11}$$

With (5.9) and (5.10), it follows that

$$\Delta\beta^{h*} \approx \frac{\eta'}{P}\sum_{p}\left(D^p - f(\underline{W}^*\cdot\underline{U}^{*p})\right)f'(\underline{W}^*\cdot\underline{U}^{*p})\,\alpha^{\nu*p} - \frac{\alpha^{ATT*}\,\underline{U}^{F,constant*}.\Delta\left(\underline{W}^{\varepsilon*}\right)_{W\varepsilon}}{\alpha^{bias*2}+|\underline{U}^{F,constant*}|^2+\alpha^{ATT*2}+\alpha^{ATT*}\,\underline{U}^{F,constant*}.\underline{U}^{F\alpha*}} \tag{5.12}$$

$$\text{with}\quad \eta' = \eta\left(\frac{\alpha^{bias*2}+|\underline{U}^{F,constant*}|^2}{\alpha^{bias*2}+|\underline{U}^{F,constant*}|^2+\alpha^{ATT*2}+\alpha^{ATT*}\,\underline{U}^{F,constant*}.\underline{U}^{F\alpha*}}\right)$$

5.3 Degradation of learning due to coupling

To compare the performance of a neuron with coupled input signals with a coupling-free neuron, the weight vector $\underline{W}^*$ must be transformed to the equivalent weight vector $\underline{W}^{**}$, which differs from $\underline{W}$ because of the transformation of the training set due to the coupling. Comparison of the dynamics of $\underline{W}^{**}$ with $\underline{W}$ yields the degradation caused by the coupling.

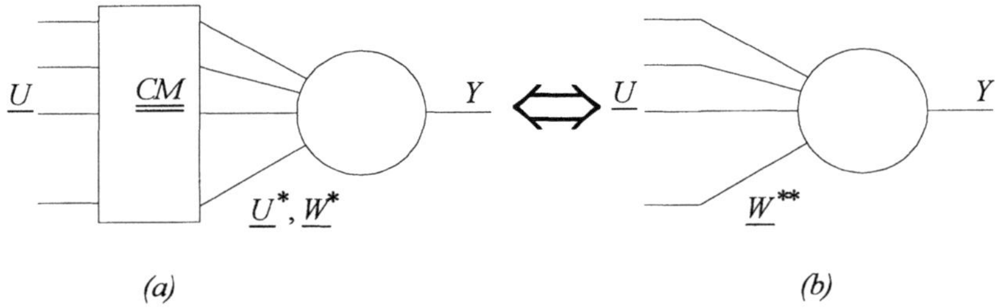

Fig. 5.3 (a) neuron with coupled input signals and (b) equivalent coupling-free neuron

Noting that $\alpha^{h*}\beta^{h*} = \alpha^h\beta^{h**}$, it follows from (5.2) and (5.3) that the relation between the norm of the weight vector in the locally optimal ($\underline{B}^h$) direction of the actual neuron and the equivalent coupling-free neuron is given by

$$\beta^{h**} = \frac{1}{|\underline{\underline{CM}}^{-1^T}\underline{B}^h|}\beta^{h*}. \tag{5.13}$$

For *training sets for which the centre-of-gravity* coincides with the origin of the input space, it follows from (5.2), (5.12) and (5.13) that the increment of β^h for a neuron without coupling between inputs and for β^{h**} of the equivalent of the neuron with coupling is changed according to

$$\Delta\beta^{h**} = \Delta\beta^h \cdot \frac{1}{|\underline{\underline{CM}}^{-1^T}\underline{B}^h|^2}. \tag{5.14}$$

In other words, when training a neuron with coupling between the input signals to learn a training set for which the centre-of-gravity coincides with the origin of input space, the learning speed appears to be lower than in case of zero coupling between input lines, by the second factor at the right hand side of (5.14), i.e. $|\underline{\underline{CM}}^{-1^T}\underline{B}^h|^{-2}$.

For training sets for which the centre-of-gravity does not coincide with the origin of input space, the effect of coupling can be derived by incorporating the effect of the slanting of the training set on the position of the centre-of-gravity in input space. As indicated by (3.36), translating the centre-of-gravity of a training set in input space changes the learning behavior. It can be concluded that in general, coupling changes the learning behavior because of *both* translating the training set in input space (see equ.(3.36)) and becase of pure slanting of the set (equ. (5.14)).

5.4 Types of coupling

In this section, a number of different types of coupling are discussed. For each of the types, some assumptions on the elements of the coupling matrix $\underline{\underline{CM}}$ are presented.

5.4.1 Capacitive coupling

For capacitive coupling between two input lines, the mutual input coupling is proportional to the difference between the two signals and proportional to the frequency of the difference-signal. Assuming that the frequency of the difference-signal of two lines is in a certain relatively small frequency band, the coupling between neighboring signal lines is approximately constant. Denoting the coupling factor between neighboring signal lines as *cm*, the ith input signal of the neuron becomes

$$\underline{U}_i^* = \underline{U}_i + cm \cdot \left(\underline{u}_{i-1} - \underline{u}_i\right) + cm \cdot \left(\underline{u}_{i+1} - \underline{u}_i\right), \tag{5.15}$$

where $\underline{u}_i$ denotes the small signal part of $\underline{U}_i$. It follows that the adaptation of the virtual equivalent weight vector in the local optimal direction, $\Delta\underline{\beta}^{h**}$, is generally not equal to the weight vector adaptation in the local optimal direction for a cross-talk-free neuron $\Delta\underline{\beta}^h$ because of *two* effects. Firstly, the training set is slanted due to the signal-coupling resulting in an increased training time required to reach a certain level of performance, see (5.14). Secondly, the adaptation is generally modified because the training set will most probably be translated in input space, which situation is included in (5.12).

5.4.2 Resistive coupling

Resistive coupling between input signals may occur due for example to non-ideal power supplies (including effects due to the resistive substrate). With a coupling factor between neighboring signal lines equal to *cm*, the ith input signal of the neuron is

$$\underline{U}_i^* = \underline{U}_i + cm\left(\underline{U}_{i-1} - \underline{U}_i\right) + cm\left(\underline{U}_{i+1} - \underline{U}_i\right). \tag{5.16}$$

In this case the ratio between $\Delta\underline{\beta}^{h**}$ and $\Delta\underline{\beta}^h$ is also modified because of both slanting the training set and because of generally translating the training in input space. Only in case of coinciding centre-of-gravity and origin of input space for the unslanted training set, the centre-of-gravity will not be translated in input space.

5.4.3 Additive coupling

An additive type of coupling can be caused by for example AM multiplex systems used to decrease the number of input-output pins of the neural network [1],[2]. With a coupling factor between neighboring signal lines equal to *cm*, the ith input signal of the neuron is

$$\underline{U}_i = \underline{U}_i + cm\,\underline{U}_{i-1} + cm\underline{U}_{i+1}. \tag{5.17}$$

Also in this case, the ratio between $\Delta\underline{\beta}^{h**}$ and $\Delta\underline{\beta}^h$ is modified due to both slanting the training set and due to generally translating the training in input space.

5.5 Calculation & simulation results

In this section, results from calculations and simulations on the increment of training time due to coupling between input signals of neurons are presented. These results were obtained *under condition* that the coupling of input signals does not change the position of the centre-of-gravity in input space in order to see exclusively the effect of slanting the training set. Situations for which both the training set is transformed due to coupling and for which simultaneously the centre-of-gravity is translated in input space will not be analyzed because they are merely a combination of slanting and separately translating the centre-of-gravity. The calculations and simulations have been done using a symmetric coupling-matrix which means that the mutual coupling between any neighboring lines is assumed identical.

Resistive and capacitive coupling

For resistive and capacitive coupling, the coupling between adjacent signals is given by (5.15) and (5.16). Assuming a stationary centre-of-gravity, the effect of resistive and capacitive coupling is identical. The coupling matrix that has been used in the analysis for both resistive and capacitive coupling is

$$\underline{\underline{CM}}(cm, N_{in}) = \begin{bmatrix} 1-cm & cm & 0 & 0 & \dots & 0 \\ cm & 1-2cm & cm & 0 & \dots & \dots \\ 0 & cm & \dots & \dots & 0 & 0 \\ 0 & 0 & \dots & \dots & cm & 0 \\ \dots & \dots & 0 & cm & 1-2cm & cm \\ 0 & \dots & 0 & 0 & cm & 1-cm \end{bmatrix}, \tag{5.18}$$

where cm denotes the coupling between neighboring signal lines, and N_{in} is the number of input lines of the neuron (excluding the bias input line).

Figure 5.4a shows the calculated decrease of the adaptation of the weight vector of the neuron in the local optimal direction as a function of the coupling factor cm in (5.18) for three numbers of input signals. It can be derived from figure 4a that the adaptation of β^{h**} is heavily dependent on the coupling factor cm.

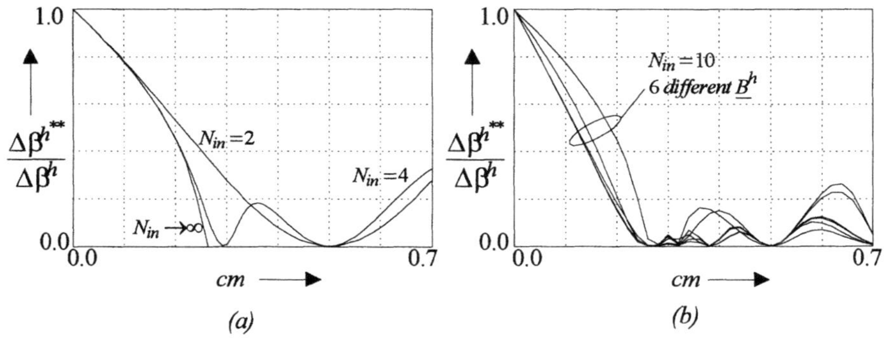

Fig. 5.4 Decrement of the adaptation in the local optimal direction related to coupling-free neuron as a function of cm (resistive and capacitive coupling); (a) for different N_{in} and (b) for $N_{in}=10$ and six different $\underline{B}^h$

It follows from (5.14) that the decrement of the adaptation of β^{h**} is dependent on the direction of $\underline{B}^h$, which dependency of $\Delta\beta^{h**}/\Delta\beta^h$ is illustrated in figure 5.4b for a neuron with a 10-dimensional input vector (excluding the bias input) for 6 different $\underline{B}^h$. Furthermore, the dependency of this ratio on $\underline{B}^h$ increases with the coupling.
The ratio at which the training time required to reach a certain level of performance increases, is the inverse of the ratio $\Delta\beta^{h**}/\Delta\beta^h$ (see chapter 3). For capacitive and resistive coupling, the coupling factor is usually smaller than 0.1, from which it can be concluded that generally no special precautions are required to decrease coupling.

Additive coupling
The coupling matrix for additive coupling is given by (5.19); the elements in this matrix result directly from (5.17). The unity elements on the diagonal of the coupling-matrix represent direct coupling from the vector $\underline{U}$ to the input of the neuron $\underline{U}^*$. The coupling between neighboring signal lines is modeled by a non-zero coupling factor *cm* between any neighboring input signals except for the bias input signal. Denoting the coupling matrix as $\underline{\underline{CM}}(cm, N_{in})$ where *cm* denotes the coupling between adjacent signal lines and N_{in} is the number of input signals of the neuron,

$$\underline{\underline{CM}}(cm, N_{in}) = \begin{bmatrix} 1 & cm & 0 & 0 & \ldots \\ cm & 1 & cm & 0 & \ldots \\ 0 & cm & 1 & cm & \ldots \\ 0 & 0 & cm & 1 & \ldots \\ \ldots & \ldots & \ldots & \ldots & \ldots \end{bmatrix} \tag{5.19}$$

Figure 5.5a presents the calculated decrease of the weight adaptation in the local optimal direction for three number of input signals as a function of the coupling between adjacent signals. Simulation results for $N_{in}=2$ and $N_{in}=4$ illustrate the correspondence with simulations. From simulations and calculations it appears that the ratio between β^{h**} and β^{h*} is independent of the direction of the $\underline{B}^h$ vector component for small coupling levels only. Figure 5.5b shows this ratio as a function of the coupling factor for a neuron with a 10-dimensional input vector for 6 different $\underline{B}^h$ vector components. The ratio $\Delta\beta^{h**}/\Delta\beta^h$ appears to be very $\underline{B}^h$-dependent for coupling factors higher than approximately 0.3.

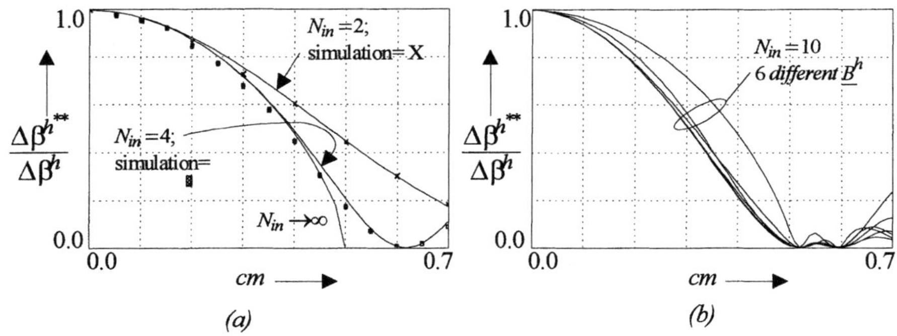

Fig. 5.5 Decrement of the adaptation in the local optimal direction related to coupling-free neuron as a function of cm (additive coupling); (a) for different N_{in} and (b) for $N_{in}=10$ and six different $\underline{B}^h$

For additive coupling (coupling that occurs in for example multiplex systems) the coupling is reported [1] in the range 0.0 to 0.6. Because low coupling factors put high demands on the modulators and demodulators in the multiplexer, low coupling factors are more difficult to realize than higher coupling factors. It follows from figures 5.5a and 5.5b that coupling factors up to about 0.3 are acceptably low for additive coupling, which corresponds to the reported tolerable coupling in [2].

5.6 Discussion

Due to coupling between input signals of a neuron, the input space is transformed. The effect of transforming the input space is that the adaptation of the weight vector in the locally optimal direction is generally decreased. The reduction in adaptation of the weight vector in the local optimal direction is due to both translating the centre-of-gravity in input space and to slanting the training set.

From the calculation and simulations it follows that the effect of slanting the training set due to resistive or capacitive coupling between input signals on the adaptation of the weight vector in the local optimal direction is small for coupling factors up to about 0.1. Noting resistive and capacitive coupling factors in the order of 0.1 are relatively high for on-chip hardware realizations, it can be concluded that no special care needs to be taken to decrease resistive and capacitive coupling between input signals. The effect of the generally translated centre-of-gravity due to coupling has been discussed in the section 5.2.

The additive coupling that occurs in for example AM-multiplex systems results in less slanting of the training set for identical coupling factors between adjacent signals. It follows from the calculations and simulations that coupling factors up to 0.3 are low enough to ensure no significant degradation of the learning behavior of the neuron. This maximum tolerable coupling, however, puts constraints on for example the AM-multiplex systems.

5.7 References

[1] M.P.Craven, K.M.Curtis and B.R.Hayes-Gill, "Frequency division multiplexing in analogue neural network", *Electronics Letters*, vol. 27, 918-920, 1991

[2] K.M.Curtis, M.P.Craven, B.R.Hayes-Gill and M.Liu, "An Optimised VLSI Implementation for Analogue Neural Communications", In *Proceedings of the 2nd International Conference on Microelectronics for Neural Networks*, Munich: Kyrill & Method Verlag, 273-279, 1990

[3] R.Gregorian and G.C.Themes, *Analog MOS Integrated Circuits for Signal Processing*, New York: Wiley, 1986

[4] C.R.Paul, "Literal solutions for time-domain crosstalk on lossless transmission lines", *IEEE Transactions on Electromagnetic Compatibility*, vol. 34, 433-444, 1992

[5] T.Sakurai, "Closed-form expressions for interconnection delay, coupling, and crosstalk in VLSI", *IEEE Transactions on Electron Devices*, vol. 40, 118-124, 1993

Chapter 6

Precision Requirements for Analog Weight Adaptation Circuitry for Single-Layer Nets

6.1 Introduction

When implementing neural networks with learning capability in analog hardware, parasitic weight adaptation due to offsets, leakage and charge injection may occur. These effects can be modeled as an extra constant weight adaptation component for each adaptation cycle. An analysis is given of the effect of constant weight adaptation on the learning behavior in sections 6.3 and 6.4. A worst case estimation of the maximum tolerable constant weight adaptation is given in section 6.5.

In the last section of this chapter, the analyses are used to estimate the worst-case minimum size for the weight storage capacitor required to properly learn a training set. It follows that this size is inversely proportional to the accuracy of the weight adaptation circuitry. Hence a trade-of between capacitor size and weight adaptation circuitry-size follows.

6.2 The cause and the model of analog imprecision

In analog on-chip learning feed-forward neural nets, analog circuitry takes care of the adaptation of every weight. Because the weights are usually stored as voltages across capacitors, the adaptation circuitry is typically an analog multiplier with charge output. This charge is usually constructed by gating an output current I^{out} during some predefined interval (τ) to the weight storage capacitor C^{weight} [1-3],[5].

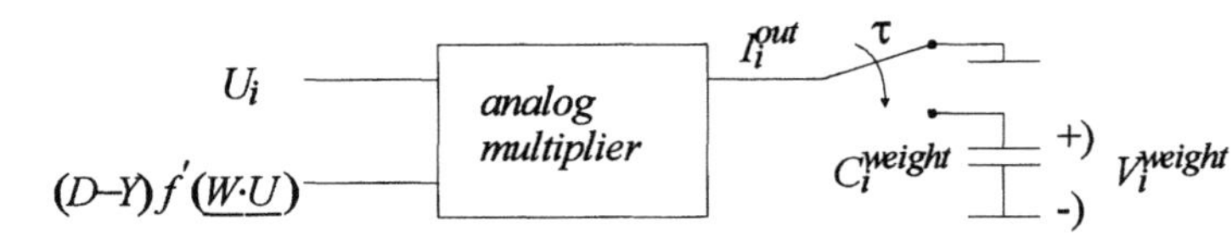

Fig. 6.1 Analog hardware realization of weight storage and adaptation circuit for i^{th} weight

The weight adaptation is the sum of the ideally wanted weight adaptation and a non-ideal (or parasitic) weight adaptation:

$$\Delta W^{total} = \Delta W + \Delta W^{par} \tag{6.1}$$

The parasitic weight adaptation is generally caused by offsets in the multiplier, offsets in its input signals, leakage of the stored weight-voltage, or is caused by charge injection of the switch that gates the output current of the multiplier to the capacitor [3]:

$$\Delta W^{par} = \Delta W^{offsets} + \Delta W^{leakage} + \Delta W^{charge\_injection} + \ldots$$

The parasitic weight adaptation due to *offsets* and due to *charge injection* is approximately proportional to the number of adaptations for the particular weight because these effects occur only during adaptation. The *leakage* effect is constant in time and therefore independent of the number of weight adaptations. However, it is assumed that the effect of leakage is small compared to the summed effects of the offsets and of the charge-injection. It follows that the *parasitic weight adaptation is by approximation constant for each weight adaptation.*

6.3 Estimation of MSE-increment due to imprecision

In this section, an analysis of the effect of constant weight adaptation on the learning behavior of single-layer feed-forward networks is presented for a relatively simple case: it is assumed for simplicity reasons that the effect of the $\underline{U}^F$ vector components on training is negligibly small. In chapter 3, this assumption leads towards the condition $|\underline{U}^{F,const}| = 0$. In a loose way , this condition means that the average (over the training examples) of any element of the $\underline{U}^F$ vector is zero. As a result of this, it has been derived in chapter 3 that the weight vector follows during learning a straight path in weight space from an initially point (usually assumed to be close to the origin of the weight space) towards the eventual spot. As a direct result of this assumption, the second term on the right hand side of (3.36) is zero.

6.3.1 Basic analysis

The parasitic adaptations of the weights form a parasitic adaptation vector $\Delta\underline{W}^{par}$, which is decomposed into three vector components related to the attractor hyperplane of the neuron:

$$\Delta\underline{W}^{par} = \beta^{h,par}\underline{B}^h + \beta^{bias,par}\underline{B}^{bias} + \underline{U}^{F,par} \tag{6.2}$$

With the equations that describe the adaptation of β^h and β^{bias}, it can be derived that the adaptation of the β^h is given by (see appendix 6.1)

$$\Delta\beta^{h,total} \approx \frac{\alpha^{bias^2}}{\alpha^{bias^2} + \alpha^{ATT^2}} \left\{ \beta^{h,par} - \frac{\beta^{bias,par}\alpha^{ATT}}{\alpha^{bias}} + \frac{\eta}{P}\sum_{p=1}^{P}\left(D^p - f(\underline{W}\cdot\underline{U}^p)\right) f'(\underline{W}\cdot\underline{U}^p)\alpha^{h,\nu,p} \right\} \tag{6.3}$$

If the neural network has reached a minimum during training (either a global or a local minimum), the net weight adaptation for any weight is ideally zero. It follows from (6.3) that if the sum of the first two terms between brackets is positive, there is a positive net adaptation of β^h in the minimum. Because the *MSE* of is a function of β^h of all neurons in a neural network (see figure 6.2), the *MSE* versus time curve has a minimum during training for this case; after reaching the minimum, the *MSE* will increase during continued training.

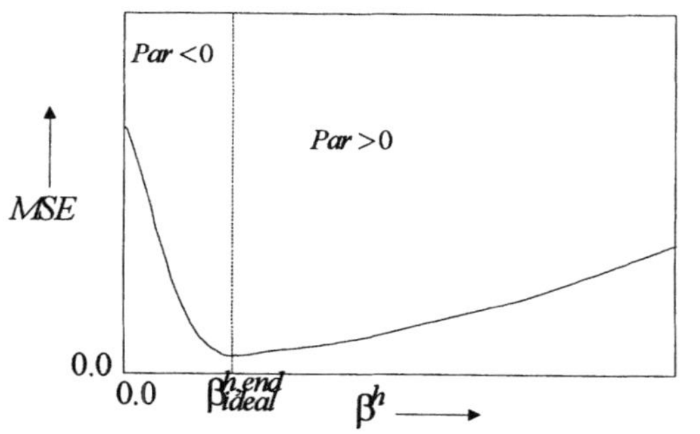

$$Par = \beta^{h,par} - \frac{\beta^{bias,par}\alpha^{ATT}}{\alpha^{bias}}$$

Fig. 6.2 Simulated MSE as a function of the β^h for a non-linearly separable training set (assuming a small $\underline{W}^{\epsilon}$); dotted line corresponds to the (global or local) minimum

6.3.2 The effect on the *MSE*

The difference between the minimum attainable *MSE* for an ideal neural net and the *MSE* which is reached by the neural net with constant weight adaptation can be approximated by

$$\begin{aligned}\Delta MSE &\approx \frac{\partial MSE}{\partial \beta^h}\cdot \Delta\beta^h \\ &\approx \frac{\beta^{h,end}-\beta^{h,end}_{ideal}}{2}\cdot \left.\frac{\partial MSE}{\partial \beta^h}\right|_{\beta^{h,end}} \\ &\approx \frac{\beta^{h,end}-\beta^{h,end}_{ideal}}{2}\left[\left.\frac{\partial MSE(\beta^h)}{\partial \beta^h}\right|_{\beta^{h,end}_{ideal}}+\left.\frac{\partial MSE}{\partial \beta^h}\right|_{\beta^{h,end}}\right]\end{aligned} \tag{6.4}$$

where:

$\beta^{h,end}_{ideal}$ is the norm of the weight vector (excluding the bias-related weight) corresponding to the global minimum.

$\beta^{h,end}$ is the norm of the weight vector (excluding the bias-related weight) at the end of training for a network with non-idealities as modelled in (6.1)

Note that because the *MSE* has a minimum for $\beta^h=\beta^{h,end}_{ideal}$ (under assumption of a negligibly small $\underline{W}^{\varepsilon}$), it can be concluded that the first derivative between brackets in (6.4) equals zero. The difference $(\beta^{h,end}-\beta^{h,end}_{ideal})$ in the previous equation can be obtained from

$$\frac{\partial^2 MSE}{\partial \beta^{h^2}}\left(\beta^{h,end}-\beta^{h,end}_{ideal}\right)\approx \left.\frac{\partial MSE(\beta^h)}{\partial \beta^h}\right|_{\beta^{h,end}}-\left.\frac{\partial MSE(\beta^h)}{\partial \beta^h}\right|_{\beta^{h,end}_{ideal}}.$$

Substituting this relation in (6.4), and noting that the right hand side of (3.36) is identical to $-\eta$ times the first derivative of the *MSE* with respect to β^h [4], it follows that under the assumption of $\underline{U}^{F,const}=\underline{0}$

$$\Delta MSE \approx \frac{\left(\beta^{h,par}-\beta^{bias,par}\dfrac{\alpha^{ATT}}{\alpha^{bias}}\right)^2}{2\,\eta^2\dfrac{\partial^2 MSE}{\partial \beta^{h^2}}} \tag{6.5}$$

Using a Taylor series expansion, it is straight forward to show that the second derivative of the *MSE* with respect to β^h is given by

$$\frac{\partial^2 MSE}{\partial \beta^{h^2}}\approx \frac{-2}{1-P(correct)}\int \varphi(D-Y)\left[(D-Y)f''(\underline{W}\cdot\underline{U})-f'^2(\underline{W}\cdot\underline{U})\right]\alpha^{h,v}\, d(D-Y)$$

In this approximation, a linearization of the derivative of the cost function E with respect to β^h is used. As a direct consequence of this, the resulting expression is therefore valid only for a limited range. Allowing only relatively small increments ΔMSE in (6.5), one does generally operate in this limited validity range. In (6.5), the distribution function for $(D-Y)$ must be known; this distribution function is determined by the total training set and by the total weight vector. A sufficiently accurate estimation of the error distribution function is calculated in appendix 6.2.

6.3.3 An illustration

Figure 6.4 illustrates the correspondence between (6.5) and simulation results for a specific training set. For the calculations, the approximation of the error distribution function that is described in appendix 6.2 was used.

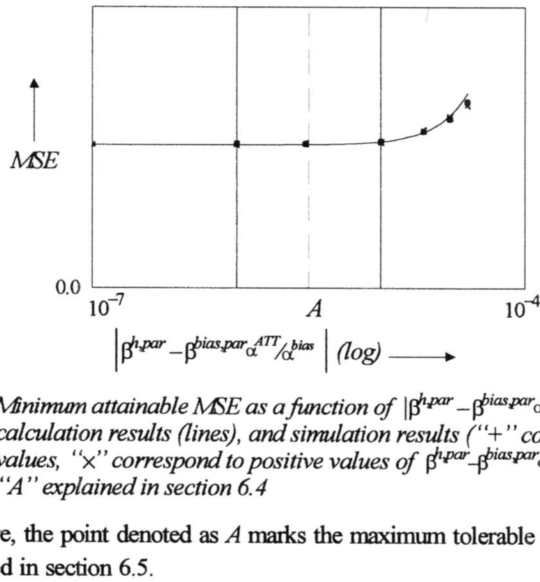

Fig. 6.3 *Minimum attainable MSE as a function of* $|\beta^{h,par} - \beta^{bias,par}\alpha^{ATT}/\alpha^{bias}|$; *calculation results (lines), and simulation results ("+" correspond to negative values, "x" correspond to positive values of* $\beta^{h,par} - \beta^{bias,par}\alpha^{ATT}/\alpha^{bias}$); *"A" explained in section 6.4*

In the figure, the point denoted as A marks the maximum tolerable constant weight adaptation as calculated in section 6.5.

6.4 The effect on correctly classified examples

It has been shown in section 6.3 that if parasitic weight adaptation as modelled in (6.1) is non-zero, then $\beta^{h,end} \neq \beta^{h,end}_{ideal}$. Section 6.3 showed furthermore that due to this being non-ideal of the $\beta^{h,end}$ (in case of non-zero parasitic weight adaptation), the *MSE* of the neural network is non-minimum. In this section, the effect of $\beta^{h,end} \neq \beta^{h,end}_{ideal}$ on the fraction of correctly classified training examples is analyzed.

Analysis

It has been mentioned in among other chapter 3 that for pattern classification application, we implictly assume that the double threshold variant of the back-propagation algorithm is used. For this double threshold variant, the desired responses D^p are either "low" or "high", and the classification of examples is assumed to be correct for network responses Y^p that satisfy

$$Y^p \geq D^p \big\backslash_{D^p = D^{high}}$$

and

$$Y^p \leq D^p \big\backslash_{D^p = D^{low}} .$$

Figure 6.4 shows an illustration of classification using the double threshold variant of back-propagation. For the two-dimensional non-linearly separable training set (note that for illustration reasons, $|\underline{U}^{F,const}| \neq 0$), the target response of examples out of $class_0$ is D_0 and the desired response for $class_1$ examples is D_1. Note that because the desired responses for both classes are constant, all vectors for which $Y=D$ form a hyperplane in parallel to *the* hyperplane of the neuron (i.e. the hyperplane that corresponds to $\underline{W} \cdot \underline{U} = 0$).

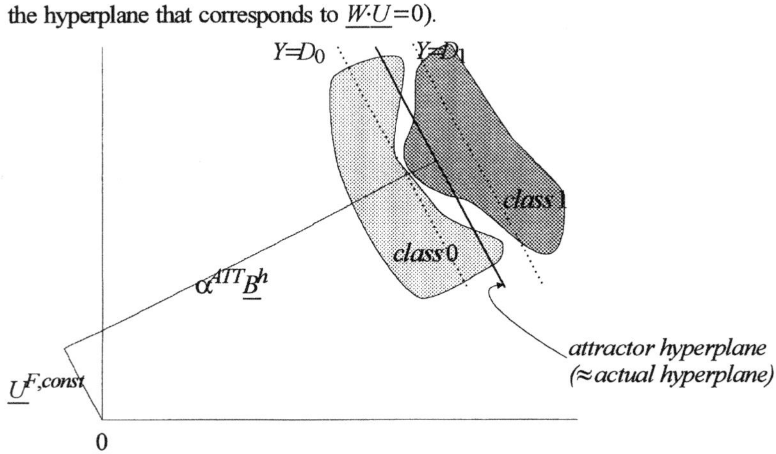

Fig. 6.4 Two dimensional training set in the input space; the two classes are denoted by "class 0" and "class 1"

These two *threshold hyperplanes* are dotted in figure 6.4. It is straight forward to derive the distance (in input space) between the two hyperplanes that correspond to the desired responses D_0 and D_1 by noting that the distance between an example and the attractor hyperplanes is

$$\alpha^{h,\nu} = \frac{f^{-1}(Y)}{\beta^h}: \tag{6.6}$$

the distance between the two threshold hyperplanes (for which $Y=D_0$ or $Y=D_1$) is then

$$\Delta\alpha^h = \frac{f^{-1}(D_1) - f^{-1}(D_0)}{\beta^h}.$$

With this equation, it follows that the distance between the two threshold hyperplanes is proportional to the reciprocal value of β^h. As a direct result, the number of not-correctly classified training examples decreases with increasing β^h:

$$P(correct) = \kappa(\beta^h) \tag{6.7a}$$

with

$$\frac{\partial\kappa(\beta^h)}{\partial\beta^h} > 0 \tag{6.7b}$$

Note that this is due to the fact that by increasing the β^h, only the fraction of correctly classified *linearly separable* examples increases whereas the fraction of correctly classified *not-linearly separable* examples remains zero. Hence, (6.7) can be subdivided into two parts, one part representing the linearly separable examples and one part representing the non-linearly separable examples:

$$\begin{aligned} P(correct) &= P(correct \mid lin\_sep)\,P(lin\_sep) + P(correct \mid non\_lin\_sep)\,P(non\_lin\_sep) \\ &= K_{lin\_sep}(\beta^h) + K_{non\_lin\_sep}(\beta^h) \end{aligned} \tag{6.8a}$$

with

$$\frac{\partial K_{lin\_sep}(\beta^h)}{\partial\beta^h} > 0 \tag{6.8b}$$

$$\frac{\partial K_{non\_lin\_sep}(\beta^h)}{\partial\beta^h} = 0 \tag{6.8c}$$

The next section of this chapter presents an estimation on the maximum tolerable parasitic weight adaptation, ensuring that the eventual performance of the neuron (or single-layer neural network) is sufficiently close to the optimal performance.

6.5 Rule of thumb

It can be derived that in case that the attractor hyperplane and the actual hyperplane given by $\underline{W}\cdot\underline{U}=0$ are close, the average adaptation of β^h is positive only for linearly separable examples that result in a non-zero weight adaptation. Similarly that all non-linearly separable examples result in an average decrease of β^h.

$$\begin{cases} \underline{U}^{lin\_sep} & \Rightarrow \overline{\Delta\beta^h} \geq 0 \\ \underline{U}^{non\_lin\_sep} & \Rightarrow \overline{\Delta\beta^h} < 0 \end{cases} \tag{6.9}$$

In these relation corresponds the superscript *lin_sep* to examples that are linearly separable and correspond input vectors with the superscript *non_lin_sep* to non-linearly separable training examples. The relations in (6.9) can easily be explained using figure 6.5. Assume that the attractor hyperplane and the actual hyperplane are close, and that $\alpha^{ATT}=0$. For this case, examples on the right hand side of the hyperplane satisfy the condition $\alpha^h>0$, while for all examples on the left hand side of the hyperplanes $\alpha^h<0$.

The adaptation of β^h for a training example is:

$$\Delta\beta^{h,p}=\eta(D-Y)f'(\underline{W}\cdot\underline{U})\alpha^h$$

and hence the sign of the adaptation of β^h equals the sign of the product $(D-Y)\alpha^h$. It is now easy to verify that adaptation on any linearly separable example results in an increase of β^h while adaptation on any non-linearly separable example decreases β^h:

$$\Delta\beta^{h,p}\Big|_{\{lin\_sep\,,\,\alpha^{ATT}=0\}} \geq 0$$

$$\Delta\beta^{h,p}\Big|_{\{non\_lin\_sep\,,\,\alpha^{ATT}=0\}} < 0$$

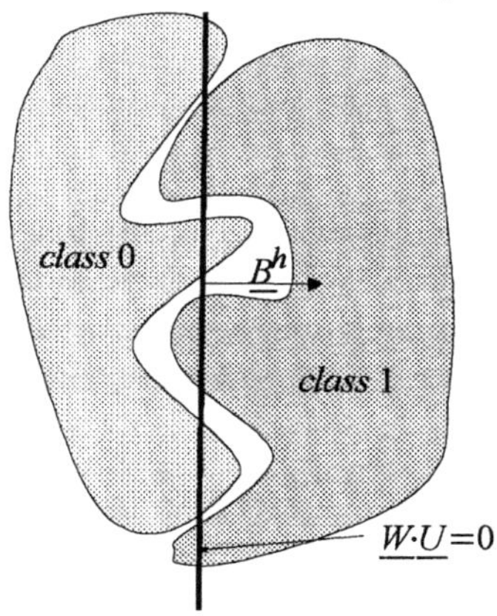

Fig. 6.5 Non-linearly separable training set

For training sets for which $\alpha^{ATT}\neq 0$, every weight adaptation has an α^{ATT}-dependent part which cancels in first order over all training examples. Because of this cancellation over all training examples, only statements concerning average adaptation over types of examples can be made. It follows that the *average adaptation* of β^h is positive for linearly separable examples and the *average adaptation* of β^h for non-linearly separable examples is negative; in formula this is given in (6.9).

6.5.1 The condition for negligibly small effect of parasitic weight adaptation

In (6.9), the average effect of linearly separable and non-linearly separable examples on the adaptation of β^h is presented. The relations in (6.9) can be interpreted as follows: in the global minimum, the linearly separable examples generate a "force" which tends to increase β^h. The non-linearly separable examples generate a "force" that tends to reduce β^h. This is illustrated in figure 6.6, which is basically figure 6.2 zoomed out around the optimum $\beta^{h,end}_{ideal}$.

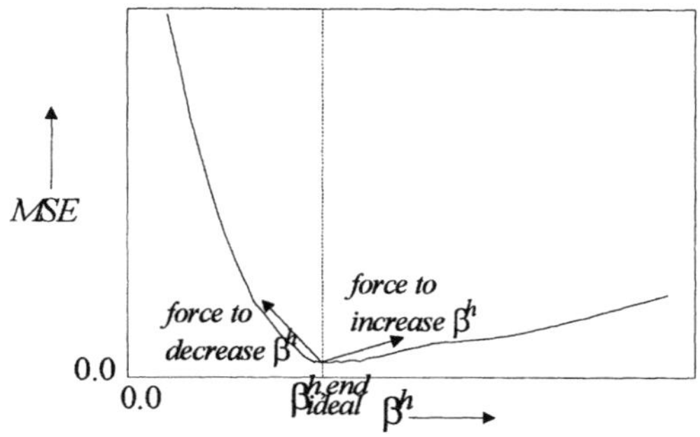

Fig. 6.6 In the minimum in the MSE-β^h curve, non-linearly separable examples generate a "force" that tends to reduce β^h; linearly separable examples tend to increase β^h

In the (global or local) minimum in the *MSE* versus β^h curve (see figure 6.2), the total adaptation of β^h is ideally zero. The total adaptation of β^h on linearly separable examples in the minimum is therefore equal to the adaptation on non-linearly separable examples but with oposite sign:

$$\overline{\Delta\beta^h(\underline{U}^{lin\_sep})}\Big\backslash_{minimum} = -\overline{\Delta\beta^h(\underline{U}^{non\_lin\_sep})}\Big\backslash_{minimum} \tag{6.10}$$

As a rule of thumb, the effect of the parasitic constant weight adaptation on the eventual performance of the neuron is negligibly small if either of the ideal adaptations of β^h in (6.10) is at least one order of magnitude larger than the parasitic weight adaptation terms on the right hand side of (6.3). In formula:

$$\left|\beta^{h,par} - \beta^{bias,par}\frac{\alpha^{ATT}}{\alpha^{bias}}\right| \ll \eta \int_{lin\_sep} INT(D\text{-}Y)\ d(D\text{-}Y) \Leftrightarrow \textit{small effect} \tag{6.11}$$

with

$$INT(D\text{-}Y) = \mathcal{P}(D\text{-}Y)\frac{\left(D\text{-}f\left(\alpha^{h,v}(D\text{-}Y)\,\beta^h\right)\right)f'\left(\alpha^{h,v}(D\text{-}Y)\,\beta^h\right)\ \alpha^{h,v}(D\text{-}Y)}{P(correct)}$$

$\mathcal{P}(D\text{-}Y)$ approximated in section appendix 6.2.

The point marked with A in figure 6.3 corresponds to the situation in which the right hand side of (6.11) is exactly one order of magnitude larger than the left hand side of (6.11). For the training sets that were used in the simulations, the deviation from the optimum is:

- 1% for the MSE (deviations in the MSE are always increments)
- 10% for the $\beta^{h,end}$, related to the ideal $\beta^{h,end}_{ideal}$ (the difference can be either an increment or a decrement, depending on the sign of $\beta^{h,par} - \beta^{bias,par}\alpha^{ATT}/\alpha^{bias}$)

With (6.3), equation (6.11) can be rewritten into

$$\left|\beta^{h,par} - \beta^{bias,par}\frac{\alpha^{ATT}}{\alpha^{bias}}\right| \leq \frac{\eta}{10\,\beta^{h,end}_{ideal}} \int_{lin\_sep} \mathcal{P}(D-Y)\,\frac{(D-Y)\,f'\!\left(f^{-1}(Y)\right)f^{-1}(Y)}{P(correct)}\,d(D-Y). \quad (6.12)$$

Note that the integral in (6.12) requires only very little information about both the training set and about the neural network including learning. Parameters included in (6.12) are:

- the shape of the activation function $f(\cdot)$
- the double threshold values D^p, and
- the distribution of the error $(D-Y)$ (approximated in appendix 6.2; this approximation of the error distribution distribution requires only a few basic performance parameters).

It follows directly from (6.12) that the maximum tolerable constant weight adaptation decreases linearly with the reciprocal value of $\beta^{h,end}_{ideal}$. With (6.6) it follows then that the maximum tolearble constant (parasitic) weight adaptation decreases with decreasing distance between the two threshold hyperplanes (i.e. between the two boundaries at which the neuron classifies examples as $Y=D$).

Although this phenomenon may appear strange at first sight, it can be explained as follows. For large $\beta^{h,end}$, small differences between $\beta^{h,end}$ and $\beta^{h,end}_{ideal}$ have very little effect on the MSE. As a result of this, the "force" that drives β^h to $\beta^{h,end}_{ideal}$ is small and hence systematic weight adaptation errors have a relatively large effect on the eventual $\beta^{h,end}$.

6.6 Worst-case estimation of precision requirements

Assuming that the constant weight adaptation ΔW^{par} is identical for all weights, (6.12) can be used to give a *worst-case* estimation of the maximum tolerable constant weight adaptation:

$$|\Delta W^{par}| \cdot \left|\sqrt{N_{in}-1} - \frac{\alpha^{ATT}}{\alpha^{bias}}\right| \approx \frac{\eta}{10\,\beta_{ideal}^{h,end}} \int_{Lin.Sep.} \wp(D-Y)\,\frac{f'\left(f^{-1}(Y)\right)f^{-1}(Y)}{P(correct)}\,d(D-Y). \tag{6.13}$$

The α^{bias} is usually equal to the magnitude of any other element in $\underline{U}$. Denoting N_{in} as the number of inputs of the neuron, it follows that $\left|\alpha^{ATT}/\alpha^{bias}\right| < \sqrt{N_{in}-1}$ and that worst-case

$$\Delta W^{par} \le \frac{\eta}{20\sqrt{N_{in}-1}\;\beta_{ideal}^{h,end}} \int_{Lin.Sep.} \wp(D-Y)\,\frac{(D-Y)f'\left(f^{-1}(Y)\right)f^{-1}(Y)}{P(correct)}\,d(D-Y). \tag{6.14}$$

Calculations show that the integral in (6.14) is weakly dependent on the distribution of $(D-Y)$. The integral changes slightly only for extreme performance parameters (either *high* fraction of correctly classified examples and simultaneous a *high MSE* or a low *P(correct)* and simultaneously a low *MSE*).

For a sigmoid activation function for the neuron (with unity maximum derivative) and double thresholds $D^p \in \{0.05, 0.95\}$, values for the integral are typically in the range [0.01, 0.02]. With these values, a worst case estimation for the maximum tolerable constant weight adaptation is:

$$\Delta W^{par} \le \frac{\eta}{20\sqrt{N_{in}-1}\;\beta_{ideal}^{h,end}}\,0.01 \qquad \left(f(x) = \frac{1}{1+exp(-4x)},\; D \in \{0.05, 0.95\}\right). \tag{6.15}$$

For higher constant weight adaptation values than those indicated by (6.15), the neuron may learn the training set but the *MSE* then depends on both the training set (especially on α^{bias} and α^{ATT}) and on the exact values of $\beta^{h,par}$ and $\beta^{bias,par}$.

Pointers

The value of $\beta_{ideal}^{h,end}$ can be derived by assuming that the incorrect knowledge part of $\underline{W}$, $\underline{W}^{\varepsilon}$, is small after training. In this case, $\beta_{ideal}^{h,end}$ is by approximation equal to the norm of the weight vector of the neuron after training, excluding the bias-related weight; in formula:

$$\underline{W}_{ideal}^{end} = \left[\underline{W}_{ideal}^{end*} \mid W^{bias}\right]^T \quad\Rightarrow\quad \beta_{ideal}^{h,end} \approx \left\|\left[\underline{W}_{ideal}^{end*} \mid 0\right]^T\right\| \tag{6.16}$$

It has been derived in chapter 3 that the maximum usable learning rate η (in order to ensure stability during training) decreases linearly with N_{in}. With (6.15), it follows that the worst-case tolerable constant weight adaptation scales as:

$$\Delta W_{wc}^{par} \propto \left(N_{in}\sqrt{N_{in}-1}\right)^{-1} \tag{6.17}$$

6.7 Estimation of minimum weight-storage C size

In this section, the analyses of the previous sections of this chapter are used to estimate the minimum size of weight storage capacitors in single-layer neural networks. For this estimation, a few additional parameters are required:

- $[-\hat{W}, \hat{W}]$ is the hardware implemented weight range
- $[-\hat{V}^{weight}, \hat{V}^{weight}]$ is the weight voltage range (stored on the weight-storage capacitor) that corresponds to the implemented weight range.
- ΔQ^{par} is the average value of the charge injection of the weight adaptation circuit (including offset phenomena)
- C^{weight} is the size of the weight storage capacitor

It is now straight forward to show that the weight storage capacitors must satisfy

$$C_i^{weight} \geq \frac{\hat{W}\,\Delta Q^{par}}{\Delta W^{par}\;\hat{V}^{weight}}, \tag{6.18}$$

where the desired accuracy for the weight adaptation, ΔW^{par}, is given by (6.14). The largest β^h possible with the specified weight range is $\hat{W}\sqrt{N_{in}-1}$. It follows with (6.15) that for a sigmoid non-linearity with *unity maximum derivative* and double *thresholds* at 0.95 respectively 0.05 of the maximum neuron response, a worst case estimation for the size of the weight storage capacitor is

$$C_i^{weight} \geq 2000 \frac{\hat{W}^2\;\Delta Q^{par}\;(N_{in}-1)}{\eta\;\hat{V}^{weight}} \qquad \text{(worst-case)}$$

For $\Delta Q^{par} = 100\ electrons$; $\hat{V}^{weight} = 1.5$; $\hat{W} = 10$; $\eta = 0.1$ and $N_{in} = 5$, the minimum size of the weight storage capacitor is then worst-case $80pF$.

Some notes

Note that depending on the training set, the hardware neural network may learn the training set properly using smaller capacitors because among others the actual $\beta_{ideal}^{h,end}$ depends on the training set. If for example the eventual hyperplane is perpendicular to one of the axes that span the input space and at the same time only half of the weight range is used, the required capacitors need worst case to be $10pF$. Note that in this case the gain of the weight adaptation block is decreased by a factor 8 with respect to the situation with the $80pF$ weight storage capacitor in order to reach the same learning rate η (using the same circuitry).

Note furthermore that the estimations in this section are worst-case estimations. As indicated by figure 6.3, constant weight adaptations up to a factor 3 larger than those indicated by equations (6.14) and (6.15) may have a small effect on the attainable performance.

6.8 Conclusions

This sction presents a mathematical estimation of precision requirements for analog weight adaptation circuitry for single-layer feed-forward neural nets. It is shown that for a specific nonlinearity and specific threshold values, the worst-case precision depends only on the learning rate, the eventual norm of the weight vector and on the dimension of the input space. For precisions lower than those indicated by the worst-case estimations in this section, the neural network may learn the training set, but the eventual performance will be heavily dependent on both training set and on the precisions of all weight adaptation blocks. With the precision estimations in this section, one can estimate whether a given analog hardwired on-chip learning neural networks are feasible for the given training set. Another application of the analysis in this section is the estimation of the minimum size for weight storage capacitors.

6.9 References

[1] D.D.Caviglia, M.Valle and G.M.Bisio, "Effects of Weight Discretization on the Back Propagation Learning Method: Algorithm Design and Hardware realization", In *Proceeding of the International Joint Conference on Neural Networks (San Diego)*, vol. II, 631-637, 1990

[2] D.D.Caviglia, M.Valle and G.M.Bisio, "A CMOS Analog Architecture for Adaptive Neural Networks", In M.Sami and J.Cadzadilla-Daguerre (Eds.) *Silicon Architectures for Neural Nets*, 113-123, 1991

[3] D.E.Rumelhart, G.E.Hinton and R.J.Williams, "Learning internal representations by error propagation", *Parallel Distributed Processing: Exploration in the microstructure of cognition* (Vol. 1), Cambridge, MA.:MIT Press, 1986

[4] M.Valle, D.D.Caviglia and G.M.Bisio, "An Experimental Analog VLSI Neural Chip with On-Chip Back-Propagation Learning", In *Proceedings European Solid-State Circuits Conference (Copenhagen)* pp. 203-206, 1992

[5] Y.Wang, "A Modular Analog CMOS LSI for Feedforward Neural Networks with On-Chip BEP Learning", In *Proceedings of the IEEE International Symposium on Circuits and Systems (Chicago)*, 2744-2747, 1993

Appendix 6.1: Derivation of equation (6.3)

In this appendix, it will be shown that under condition of $|\underline{U}^{F,const}| = 0$, the adaptation of the β^h_{total} is given by (6.3). In the derivation, the α^h is decomposed into two components. One component corresponds to the α^h of a vector that ends in the attractor hyperplane and is denoted as α^{ATT}, this component is quasi-static. The other component has by approximation a zero-mean value and is denoted as $\alpha^{h,v}$:

$$\alpha^h = \alpha^{h,v} + \alpha^{ATT}. \tag{6A-1}$$

The adaptation of β^h and β^{bias} is now given by

$$\Delta\beta^h \approx \frac{\eta}{P}\sum_{p=1}^{P}\left(D^p - f(\underline{W}\cdot\underline{U}^p)\right)f'(\underline{W}\cdot\underline{U}^p)\alpha^{h,p} \text{ and } \Delta\beta^{bias} \approx \frac{\eta}{P}\sum_{p=1}^{P}\left(D^p - f(\underline{W}\cdot\underline{U}^p)\right)f'(\underline{W}\cdot\underline{U}^p)\alpha^{bias}.$$

Assuming that the attractor hyperplane is quasi-stationary in the input space, it follows that under condition of a negligibly small $\underline{W}^{\varepsilon}$,

$$\frac{\Delta\beta^{h,total}}{\Delta\beta^{bias,total}} \approx -\frac{\alpha^{bias}}{\alpha^{ATT}}. \tag{6A-2}$$

The total adaptation of β^h (including the parasitic constant adaptation) can now be written as

$$\Delta\beta^{h,total} \approx \frac{\eta}{P}\sum_{p=1}^{P}\left(D^p - f(\underline{W}\cdot\underline{U}^p)\right)f'(\underline{W}\cdot\underline{U}^p)\left(\alpha^{h,v,p} + \alpha^{ATT}\right) + \beta^{h,par} \tag{6A-3}$$

and the total adaptation of the bias-related weight β^{bias} (including the parasitic part) is

$$\Delta\beta^{bias,total} \approx \alpha^{bias}\frac{\eta}{P}\sum_{p=1}^{P}\left(D^p - f(\underline{W}\cdot\underline{U}^p)\right)f'(\underline{W}\cdot\underline{U}^p) + \beta^{bias,par} \tag{6A-4}$$

Substituting (6A-4) in (6A-3) it follows that

$$\Delta\beta^{h,total} \approx \frac{{\alpha^{ATT}}^2}{{\alpha^{bias}}^2}\left(\Delta\beta^h_{total} + \beta^{bias,par}\right) + \frac{\eta}{P}\sum_{p=1}^{P}\left(D^p - f(\underline{W}\cdot\underline{U}^p)\right)f'(\underline{W}\cdot\underline{U}^p)\alpha^{h,v,p} + \beta^{h,par}. \tag{6A-5}$$

With the assumption that the attractor hyperplane is stationary in input space, (6.A-2) results. Substituting (6A-2) in (6A-5) gives (6.3):

$$\Delta\beta^{h,total} \approx \frac{{\alpha^{bias}}^2}{{\alpha^{bias}}^2 + {\alpha^{ATT}}^2}\left\{\beta^{h,par} - \frac{\beta^{bias,par}\alpha^{ATT}}{\alpha^{bias}} + \frac{\eta}{P}\sum_{p=1}^{P}\left(D^p - f(\underline{W}\cdot\underline{U}^p)\right)f'(\underline{W}\cdot\underline{U}^p)\alpha^{h,v,p}\right\}$$

Appendix 6.2: Approximation of error distribution

In this chapter, the expressions use the actual error distribution $\wp(|D-Y|)$. However, it appears that approximations of the actual error-distributions are sufficiently accurate.
In this appendix, two approximations of the actual error-distribution function are presented. The first one is very simple and accurate enough for the estimations in this chapter. After this simple estimations, a more accurate approximation is presented. For estimations of the maximum tolerable constant weight adaptation, this more accurate approximation does not provide significantly more accurate estimations. However, for some other estimations (such as for example the estimation of the maximum tolerable noise-level) this more accurate estimation may be very useful.

A linear approximation

First of all, a linear approximation for the error-distribution (as a function of the error) is presented. For this distribution, it is assumed that the distribution of the error $|D-Y|$ decreases linearly with the error:

$$\wp(|D-Y|) = \begin{cases} B - A\,|D-Y| & |D-Y| < B/A \\ 0 & elsewhere \end{cases},$$

the parameters A and B are estimated by solving the following two equations:

$$\begin{cases} 1 - P(correct) = \int \wp(|\varepsilon|)\, d\varepsilon \\ MSE = \int \varepsilon^2\, \wp(|\varepsilon|)\, d\varepsilon \end{cases}$$

which yiels:

$$A = \frac{(1-P(correct))^2}{3 \cdot MSE} \quad \text{and} \quad B = \sqrt{\frac{2(1-P(correct))^3}{3\,MSE}}\,.$$

An illustration of the approximation of the assumed error distribution is presented in figure 6.7. In figure 6.7a, the linearly approximated distribution of the response of the neural net is shown for the two classes to be classified by the neural network. Note that the distribution of the responses of examples that are already correctly classified is not defined. In figure 6.7a, the approximation of the distribution of the network responses is plotted for two different sets $(MSE, P(correct))$; one set corresponds to the dotted lines and the other set corresponds to the normal lines.

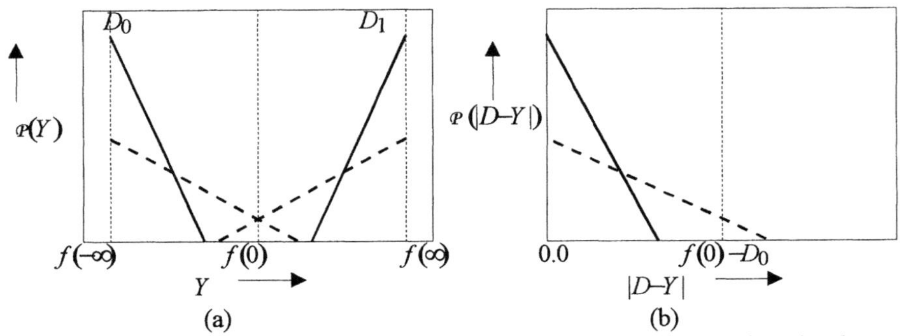

Fig. 6.7 *(a) approximation of the distribution of the response of the neural net for class$_0$ examples (with target response D_0) and for class$_1$ examples (target D_1) for two sets $(MSE, P(correct))$ and (b) the equivalent error distribution for the same two sets $(MSE, P(correct))$*

Figure 6.7b presents the equivalent distribution of the errors $|D-Y|$ as a function of the error for the same two sets $(MSE, P(correct))$. It appears that using this linear approximation of the actual error distribution does not introduce significant errors in the estimation of the maximum tolerable constant weight adaptations. This is mainly due to only using the error-distribution to estimate the first derivative of the cost-function with respect to β^h.

An exponential approximation

For applications where not only an estimation of the first derivative of the cost-function with respect to the β^h is required, but also a direct estimation on the actual size of $\Delta\beta^h$ is needed, a more accurate estimation must be used than the one in the first part of this appendix. We propose to use an exponential function to approximate the actual error-distribution function:

$$\mathcal{P}(|D-Y|) = A\exp\left(-B\cdot|D-Y|^C\right)$$

The three parameters A, B and C have to be obtained from the following equations:

$$\begin{cases} 1-P(correct) = \int \mathcal{P}(|\varepsilon|)\,d\varepsilon \\ MSE = \int \varepsilon^2\,\mathcal{P}(|\varepsilon|)\,d\varepsilon \\ \Delta\beta^h \equiv 0 = \int \varepsilon\,\mathcal{P}(|\varepsilon|)\,f'(f^{-1}(\varepsilon))\,f^{-1}(\varepsilon)\,d\varepsilon \end{cases}$$

It is clear that a major difference between the two approximations in this appendix is that for the exponential approximation more information is used; the fact that the net weight adaptation in the local or global minimum is zero has been included.

Chapter 7

Discretization of Weight Adaptations in Single-Layer Nets

7.1 Introduction

In on-chip learning neural networks, circuitry takes care of the adaptation of all weights. In the previous chapter, it was assumed that this circuitry is analog. This means that any weight adaptation can be realized. Problems concerning offsets in analog weight adaptation circuits, charge injection and leakage have been discussed and analyzed in the previous chapter.
The effect on learning of non-idealities such as offsets, charge injection and leakage are negligibly with digital weight adaptation circuits. However, with digital weight adaptation and storage, the weights are discretized, from which it is known that it has a direct effect on the learning behavior of feed-forward networks [1],[2],[4],[6],[7],[8].

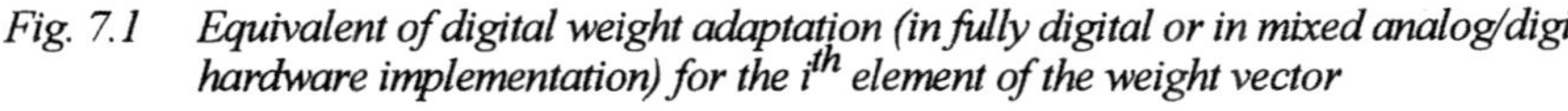

Fig. 7.1 Equivalent of digital weight adaptation (in fully digital or in mixed analog/digital hardware implementation) for the ith element of the weight vector

In this chapter, the effect of discretized weight adaptation on the learning behavior of single-layer feed-forward neural networks is analyzed. In the analysis, it is assumed that the quantization of weight adaptations is dominant, which results in a model of the weight adaptation block as shown in figure 7.1.

The analyses in the first parts of this chapter are used in section 7.4 to estimate the requirements on *both* the required resolution of digital weight adaptation circuits, *and* on weight storage resolution *during* training. Furthermore, as charge on capacitors is quantized (the stored charge is an integer multiple of the electron charge), the absolute minimum size of the weight-storage capacitor can be estimated. This estimation is presented in section 7.5.

7.2 Basics of discretized weight adaptations

Defining a quantization function $Q(x,q)$ that quantizes a value x in multiples of the quantization step size q, the adaptation of the ith element of the weight vector is

$$\Delta W_i(q) \approx \int \mathcal{P}(D{-}Y) \frac{Q\left[\eta\,(D{-}f(W{\cdot}U))\,f'(W{\cdot}U)\,U_i\,,q\right]}{P(correct)}\,dU_i\,. \tag{7.1}$$

Due to the quantization of the weight adaptation, any ideal (non-quantized) weight is hence not equal to the actual (quantized) weight adaptation. As a direct result of this, the average adaptation of any weight over all training examples generally changes due to the quantization:

$$\Delta W_i(q) = \Delta W_i + \Delta W_i^{error}(q) \tag{7.2}$$

The average difference between the ideal weight adaptation ΔW_i and the discretized adaptation $\Delta W_i(q)$ can be calculated by defining a "*quantization error function*"

$$Q\varepsilon(x,q) = Q(x,q) - x. \tag{7.3}$$

The $\Delta W_i^{error}(q)$ term in (7.2) is now

$$\Delta W_i^{error}(q) \approx \int \mathcal{P}(D{-}Y) \frac{Q\varepsilon\left[\eta\,(D{-}f(W{\cdot}U))\,f'(W{\cdot}U)\,U_i\,,q\right]}{P(correct)}\,dU_i\,. \tag{7.4}$$

In the following sections of this chapter, first a number of analyses of the effect of quantized weight adaptation on the learning behavior of signle-layer nets presented.

In section 7.3, an asymptotical approximation of the relation between the quantization step size q and the resulting $\beta^{h,end}$ is given. This asymptotical approximation can be used to estimate the *absolute maximum quantization step* q allowed during training in order to reach a certain level of performance. This estimation assumes that the performance of the neural network is maximum for a certain range q and *only* outside this range decrease with increasing quantization steps q. In a more realistic model, the performance of the neural network starts to decrease significantly *before* the absolute maximum quantization step is reached. Section 7.4 presents estimations of the maximum tolerable quantization step in order to have a negligibly small effect on training.

7.3 Performance versus quantization: asymptotical

Under assumption of stability during training (i.e under assumption of a sufficiently small η), there exists a weight vector for a neuron in a neural network that corresponds to a (local or global) minimum of the cost function E. With the VDM, the norm of the component of this weight vector that is in the $\underline{B}^h$ direction has been denoted as $\beta^{h,end}_{ideal}$ in section 6.3.

Due to quantization of weight adaptations, a neuron may not be able to reach this ideal $\beta^{h,end}_{ideal}$, but reaches the sub-optimal $\beta^{h,end}$. In this section, asymptotical approximations of the relation between the maximum attainable $\beta^{h,end}$ and the quantization step size q are presented. With the relation between the MSE and $\beta^{h,end}$ (which has been introduced in section 6.3.2), a relation between the minimum attainable MSE and the quantization step q can be derived.

In section 7.3.1, first the relation between $\beta^{h,end}$ and q for a relatively simple case will be analyzed. Sections 7.3.2 and 7.3.3 present analyses of more general cases.

7.3.1 A simple case

For simple training sets, giving an estimate for the maximum quantization size that is allowed to reach a specific $\beta^{h,end}$ is relatively simple. In this context, a "simple" training set is a training set for which the $\underline{B}^h$ unit vector has only one non-zero element and for which $\alpha^{ATT}=0$. Note that the first condition means that the attractor hyerplane is perpendicular to one of the axes that span the input space, whereas the second condition implies that the attractor hyperplane includes the origin of the input space. A training set that satisfies the conditions of being "simple" is shown in figure 7.2.

It straight forward to show that the absolute maximum attainable $\beta^{h,end}$ for quantized weight adaptations corresponds to the value of β^h for which *any* ideally positive adaptation $\Delta\beta^{h,p}$ is (at least maginally) rounded to zero.

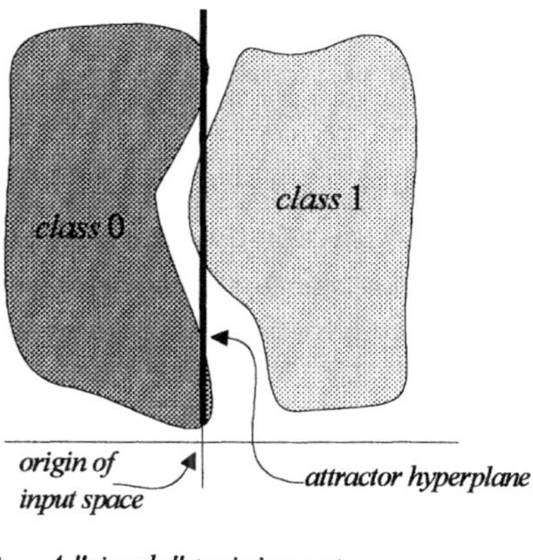

Fig. 7.2 A "simple" training set as defined in this section

This condition on the maximum attainbale $\beta^{h,end}$ can be rewritten into an expression that gives an estimation on the maximum tolerable quantization step size q in order to obtain a certain $\beta^{h,end}$ after training:

$$\left\{ q \in \mathbb{R}^+ \mid sup\left(\left|\eta(D-f(\alpha^h\beta^{h,end}))f'(\alpha^h\beta^{h,end})\alpha^h\right|\right) \geq \frac{q}{2} \right\} \tag{7.5}$$

where

the $sup(\cdot)$ operator calculates the maximum over all training examples.

it is assumed that the effect of the inproduct $\underline{W}^{\varepsilon}\cdot\underline{U}^{F}$ is small

Equation (7.5) can be rewritten into

$$\left\{ q \in \mathbb{R}^{+} \middle| \ \frac{1}{\beta^{h,end}} sup\left(\eta\,(D-f\,(\alpha^{h}\beta^{h,end}))\,f'\,(\alpha^{h}\beta^{h,end})\left(\alpha^{h}\beta^{h,end}\right)\right) \geq \frac{q}{2} \right\}. \tag{7.6}$$

For this equation, it can be derived that the value of the $sup(\cdot)$ term in (7.6) does not change during training, under assumption of a sufficiently large training set. This can be illustrated as follows: by replacing $\alpha^h\beta^h$ by a dummy variable $x \in \mathbb{R}$, the maximum for the $sup(\cdot)$ term is constant and obtained for a specific x^{max}. For a sufficiently large training set, there is always at least one example for which $\alpha^h\beta^h \approx x^{max}$. It may be clear that for small training sets, the resulting value of the $sup(\cdot)$ term may fluctuate slightly.
Hence, for sufficiently large training sets (7.6) can be rewritten into

$$\begin{cases} \beta^{h,end} = \dfrac{C}{q} & q > \dfrac{\psi}{\beta^{h,end}_{ideal}} \\ \beta^{h,end} = \beta^{h,end}_{ideal} & q \leq \dfrac{\psi}{\beta^{h,end}_{ideal}} \end{cases} \tag{7.7}$$

with $\psi = 2\ sup\left(\eta\,(D-f\,(\alpha^{h}\beta^{h,end}))\,f'\,(\alpha^{h}\beta^{h,end})\left(\alpha^{h}\beta^{h,end}\right)\right)$,

where the value of ψ is by good approximation independent of the training set for large enough training sets. Note that ψ is a function of among others the target responses D, the learning rate η, and the type of non-linear function $f(\cdot)$.

An illustration

Figure 7.3 shows the simulated and calculated (using (7.7)) attainable $\beta^{h,end}$ as a function of the quantization step size q. The training set used for this illustration is non-linearly separable.

Fig. 7.3 Simulated and calculated maximum attainable $\beta^{h,end}$ as a function of q: (for a training set for which $\alpha^{ATT}=0$)

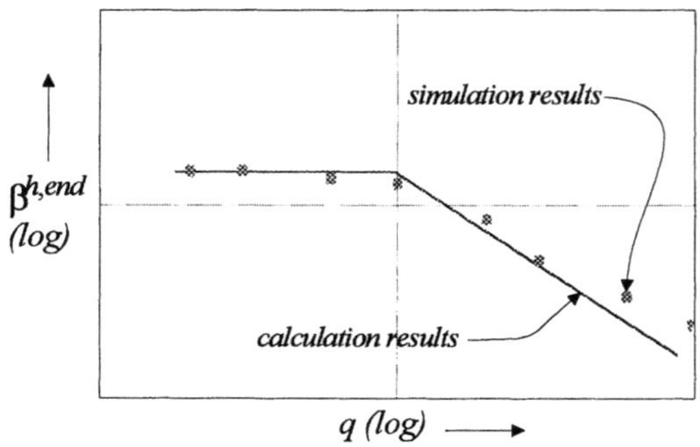

7.3.2 A less simple case

For training set for which $\alpha^{ATT} \neq 0$ but again under assumption that the $\underline{B}^h$ unit vector has only one non-zero element, calculating the maximum allowable quantization step size is more complicated than for the case in section 7.3.1. Note that for this type of training set, the attractor hyperplane is still perpendicular to one of the axes that span the input space, but the attractor hyperplane does not need to include the origin of input space.

For this type of training set, the adaptation of β^h when having a quantization step q is

$$\Delta\beta^h(q) \approx \int \mathcal{P}(D\text{–}Y)\frac{Q\left[\eta(D\text{–}f(\underline{W}\cdot\underline{U}))f'(\underline{W}\cdot\underline{U})(\alpha^{h,v}+\alpha^{ATT}),q\right]}{P(correct)}\,d\alpha^h . \tag{7.8}$$

In this case, two quantization mechanisms are present. Firstly, the mechanism described in section 7.3.1 introduces errors which may degrade the performance of the neuron after reaching $q = \Psi/\beta^{h,end}_{ideal}$. Secondly, the range of q for which the neuron may adapt correctly is expanded because of the non-zero α^{ATT}. An illustration of this second mechanism is shown in figure 7.4:

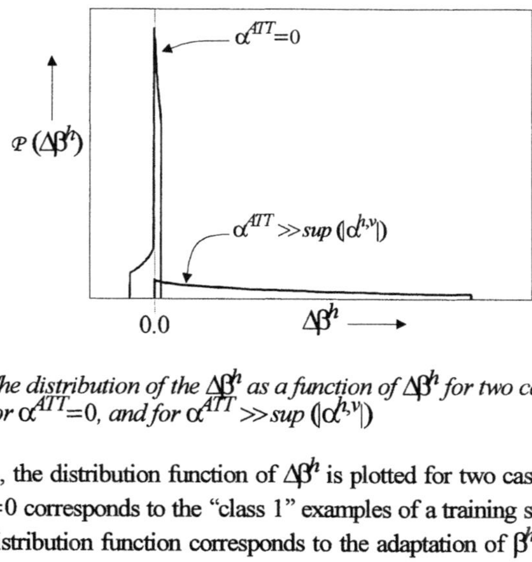

Fig. 7.4 The distribution of the $\Delta\beta^h$ as a function of $\Delta\beta^h$ for two cases: for $\alpha^{ATT}=0$, and for $\alpha^{ATT} \gg sup(|\alpha^{h,v}|)$

In figure 7.4, the distribution function of $\Delta\beta^h$ is plotted for two cases. The "spiky" distribution around $\Delta\beta^h=0$ corresponds to the "class 1" examples of a training set with $\alpha^{ATT}=0$. The lower and wider distribution function corresponds to the adaptation of β^h for the "class 1" examples of the same training set, which was translated in input space in the direction of $\underline{B}^h$, with $\alpha^{ATT} \gg sup(|\alpha^{h,v}|)$.

It follows directly from this figure that the range of adaptations $\Delta\beta^h$ increases with $|\alpha^{ATT}|$. As a direct result, larger quantization steps can be tolerated during adaptation. Note that in order to calculate at which q the performance decreases significantly, the effect of both classes must be taken into account to compensate for the α^{ATT} component in α^h (the α^{ATT} component cancels in first order for the two classes, see chapter 3).

Similar as done for the simple case, the *absolute maximum quantization step* allowed, in order to reach a specific $\beta^{h,end}$ can be approximated by

$$\left\{ q \in \mathbb{R}^+ \middle| \; sup\left(\left| \eta \, (D - f(\alpha^h \beta^{h,end})) f'(\alpha^h \beta^{h,end}) \cdot (\alpha^{h,v} + \alpha^{ATT}) \right| \right) \geq \frac{q}{2} \right\}. \tag{7.9}$$

Because in general $|\alpha^{h,v}| \ll |\alpha^{ATT}|$, this expression can be simplified into

$$\left\{ q \in \mathbb{R}^+ \middle| \; |\alpha^{ATT}| \, sup\left(\left| \eta \, (D - f(\alpha^h \beta^{h,end})) f'(\alpha^h \beta^{h,end}) \right| \right) \geq \frac{q}{2} \right\}. \tag{7.10}$$

It follows from (7.10) that under condition of $|\alpha^{ATT}| \gg sup(|\alpha^{h,v}|)$,

$$\begin{cases} \beta^{h,end} = \beta^{h,end}_{ideal} & \text{for } q \leq |\alpha^{ATT}| \, \gamma \\ \beta^{h,end} = 0 & \text{for } q > |\alpha^{ATT}| \, \gamma \end{cases} \tag{7.11}$$

$$\text{with } \gamma = 2 \; sup\left(\eta \, (D - f(\alpha^h \beta^{h,end})) f'(\alpha^h \beta^{h,end}) \right)$$

Note that equations (7.9) to (7.11) are again based on a relatively small inproduct $\underline{W}^{\varepsilon} \cdot \underline{U}^F$.

An illustration

Figure 7.5 shows the simulated and calculated $\beta^{h,end}$ as a function of q for the same training set as used for figure 7.4, but translated in input space to obtain $\alpha^{ATT} \gg sup(|\alpha^{h,v}|)$.

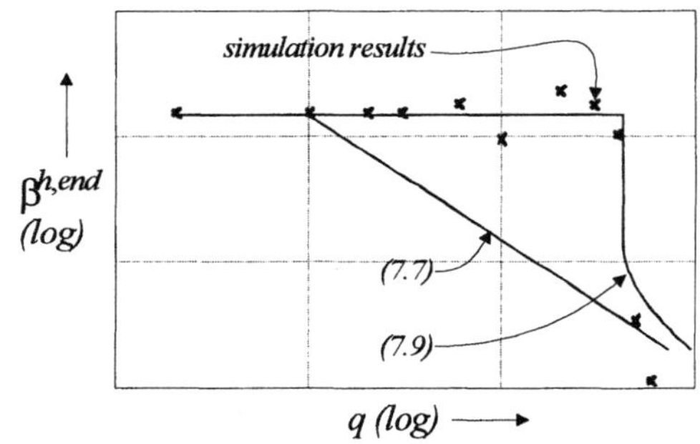

Fig. 7.5 Simulated and calculated maximum attainable $\beta^{h,end}$ as a function of q for the training set used for figure 7.4, but translated so that $\alpha^{ATT} \gg sup(|\alpha^{h,v}|)$; calculated curve according to (7.9), for comparison a curve for $\alpha^{ATT}=0$

The calculations for figure 7.5 are done using (7.9). Note that with the asymptotic approximation in (7.11) the "tail" in the calculations is not present. To clearly see the effect of the non-zero α^{ATT} on the relation between attainable $\beta^{h,end}$ and q, the calculation results of figure 7.4 have been included in the figure.

7.3.3 A general case

In general, a training set satisfies neither the condition $\alpha^{ATT} \neq 0$, nor has an $\underline{B}^h$ unit vector with exactly one non-zero elements: of the N_{in} elements in the $\underline{B}^h$ vector generally $(N_{in}-1)$ elements are non-zero. Note that by definition of the $\underline{B}^h$ vector, the N_{in}^{th} element is zero because this elements corresponds to the bias input-related weight.

In case that all elements in $\underline{B}^h$ are different, the quantization of weight adaptation generally affects all elements differently. Introducing a quantized weight adaptation vector $\underline{W}^q$ for which

$$\Delta\underline{W}^q = \underline{Q}(\Delta\underline{W}) = \left[Q(W_1), Q(W_2), \ldots, Q(W_{N_{in}})\right] \tag{7.12}$$

the quantized adaptation of $\beta^h\underline{B}^h$ is given by

$$\Delta\beta^{h*} = \underline{Q}(\Delta\underline{W}) \cdot \underline{B}^h . \tag{7.13}$$

Assuming a relatively small $\underline{W}^{\varepsilon}$ vector component, it can be assumed that the training examples that result in weight adaptation are located near to the attractor hyperplane. It follows that the distribution of the training examples in the direction of $\underline{B}^h$ is minimum and that the distribution of examples in $\underline{U}^F$ directions is maximum. This means that the distribution of the examples is minimum for directions along those axes of the input space for which the corresponding element in $\underline{B}^h$ is maximum, and vise versa.

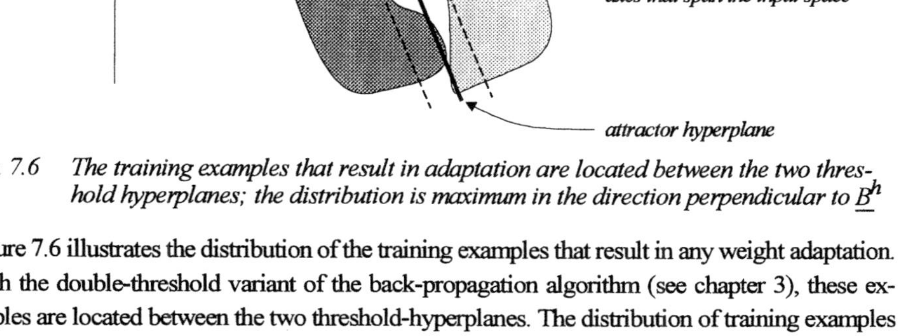

Fig. 7.6 The training examples that result in adaptation are located between the two threshold hyperplanes; the distribution is maximum in the direction perpendicular to $\underline{B}^h$

Figure 7.6 illustrates the distribution of the training examples that result in any weight adaptation. With the double-threshold variant of the back-propagation algorithm (see chapter 3), these examples are located between the two threshold-hyperplanes. The distribution of training examples that result in adaptation is hence maximum in directions perpendicular to $\underline{B}$.

As a result of this, the distribution of single *elements of the input vectors* $\underline{U}^P$ is maximum for the element U_x that corresponds to the smallest element in the $\underline{B}^h$ vector B_x^h: this element is maximally in parallel to the attractor hyperplane.

Calculation of maximum of any element of $\underline{U}$
It has been shown earlier in section 7.3 that for the calculation of the absolute maximum attainable $\beta^{h,end}$, the maximum of any element of the input vector component $\alpha^h \underline{B}^h$ is required. With the assumption that the training examples are distributed close to the attractor hyperplane, it follows that the maximum of any element of $\underline{U}$ are located near intersections of the attractor hyperplane with the edges of the training set.
Assuming that the elements of any input vector are distributed in

$$U_n \in [a\,,b],$$

the maximum of any element of $\underline{U}$ can be derived by following a monotonous ascending or descending path over the intersection between the attractor hyperplane and the edges of the training set. By noting that the attractor hyperplane is given by the equation

$$\underline{U \cdot B^h} = -\alpha^{bias}\beta^{bias}/\beta^h,$$

the maximum for the Nth element of the input vector is given by

$$max(U_N) = -\frac{\sum_{n \neq N,\, B_n^h < 0} b\,\beta^h B_n^h + \sum_{n \neq N,\, B_n^h > 0} a\,\beta^h B_n^h + \alpha^{bias}\beta^{bias}}{\beta^h B_N^h}. \tag{7.14}$$

In this equation:
the a and b may be interchanged to obtain the other extreme, and
the $max(U_N)$ must be clipped in the specified range [-a,b].
Note that for the case that $\underline{B}^h$ has exactly one non-zero element, the maximum of any element of $\underline{U}$ in directions in parallel to axes that span the input space are:

$$\begin{cases} [a\,,b] & \text{for } B_n^h = 0 \\ \alpha^{ATT} & \text{for } B_n^h = 1 \end{cases}$$

Because all elements in the summations in (7.14) have the same sign, it follows that the smallest value for $max(U_N)$ is obtained for the element that corresponds to the largest $|B_n^h|$:

$$max(U_n) = \xi(B_n^h) \tag{7.15}$$

with

$$\frac{\partial \xi(x)}{\partial x} < 0$$

Furthermore, it can easily be derived that

$$max(U_N) \geq \alpha^{ATT}. \tag{7.16}$$

In the following sub-section, an illustration of the distribution of examples that result in non-zero weight adaptation is given.

An illustration

Figure 7.7 shows the simulated distribution of training examples for a simple two-dimensional non-separable training set. The two classes to be separated are bounded by the piece-wise linear borders.

Fig. 7.7 Simulated distribution of correctly classified examples and of examples that result in a non-zero weight adaptation for W_1

The training examples which are correctly classified by the neuron are in the lower and upper corners of the training set. The separation between not-correctly classified examples and correctly classified examples is made by two hyperplanes in parallel to the actual hyperplane for which $\underline{W}\cdot\underline{U}=0$, see also figure 7.6.

The training examples for which the neuron response is not correct, are all examples inside the two classes but outside the lower and upper corner. Without quantization effects, all of the examples for which the response is not correct would result in non-zero adaptation of any element of the weight vector. However, due to quantization of weight adaptations, small weight adaptations are rounded to zero. The white areas in figure 7.7 mark the region for which all adaptation of W_1 is rounded to zero. The examples that *do* result in non-zero weight adaptations for W_1 are located in the two dotted areas close to the hyperplane.

Estimation of maximum q
From equation (7.15) it follows that smallest $max(U_n)$ occurs for the element that corresponds to the largest element in $\underline{B}^h$, B^h_n. Furthermore, it follows from (7.6) and (7.9) that the effect of quantification of weight adaptations is the least significant for elements that have the smallest value for $max(U_N)$. It can be concluded that the adaptation of the largest the element in the weight vector is the most corrupted by discretization of weight adaptations.
Worst-case, the adaptation of the weight vector in the $\underline{B}^h$ direction stops as soon as the adaptation in of the weight that corresponds to the largest element in $\underline{B}^h$ is rounded to zero for any training vector. Therefore, the maximum tolerable quantization step q in order to attain a certain $\beta^{h,end}$ is by approximation given by

$$\left\{ q \in \mathbb{R}^+ \middle| \; sup\left(\left| \eta \, (D - f(\alpha^h \beta^{h,end})) f'(\alpha^h \beta^{h,end}) \cdot max(U_N) \right| \right) \geq \frac{q}{2} \right\}, \tag{7.17}$$

where $max(U_N)$ is given in the previous equation. For an $\underline{B}^h$ with exactly one non-zero element, this expression is equal to (7.10). The effect of (7.17) on training is comparable to the effect plotted in figure 7.5.

7.4 Worst-case estimation of quantization steps

Section 7.3 presented analyses of the absolute maximum tolerable quantization step q in order to be able to obtain a certain $\beta^{h,end}$. These analyses resulted in relatively simple asymptotical relations between q and $\beta^{h,end}$. However, the asymptotical relations are not usable in estimating the maximum tolerable quantization step when requiring that the change of the level of performance due to quantization must be small. Before the point in which the asymptotical approximations in figures 7.3 and 7.5 decrease, a non-zero difference between the ideal $\Delta\beta^h$ and the quantized $\Delta\beta^h$ results.

In this section, worst-case estimations for the maximum tolerable quantization step in order to have an negligibly small effect of the quantization are given for three cases. The estimations follows the same line as sections 6.6 and 7.3.

7.4.1 A simple case

In this section, it is again assumed that the training set is relatively simple. Similar to the demands in section 7.3.1, for this simple set two conditions must be satisfied: $\alpha^{ATT}=0$ and $\underline{B}^h$ has exactly one non-zero element. Such a "simple" training set is shown in figure 7.2.

For this simple training set, the quantization on the adaptation of the $|\beta^h \underline{B}^h|$ is caused by the quantized adaptation on *one* element of the input vector. With (7.3), the average difference between ideal and actual adaptation of β^h is given by:

$$\Delta\beta^{h,error}(q) \approx \int \mathcal{P}(D-Y) \frac{Q\varepsilon\left[\eta(D-f(\underline{W}\cdot\underline{U}))f'(\underline{W}\cdot\underline{U})\,\alpha^h\,,\,q\right]}{P(correct)}\,d\alpha^h\,. \tag{7.18}$$

Note that (7.18) converts the errors in $\Delta\beta^h$ introduced by quantization into an equivalent constant weight adaptation which was analyzed in chapter 6. Now, we can define q_{crit} as the largest quantization step for which the effect of the quantization on the learning behavior is small. With the derivations in sections 6.5 and 6.6, the following expression is derived:

$$\left|\Delta\beta^{h,error}(q_{crit})\right| \approx \frac{\eta}{10\,\beta^{h,end}_{ideal}} \int_{LinSep} \mathcal{P}(D-Y)\frac{(D-Y)f'\left(f^{-1}(Y)\right)f^{-1}(Y)}{P(correct)}\,d(D-Y). \tag{7.19}$$

It follows that the effect of the quantization becomes significant for $q > q_{crit}$. Note that the effect of $\Delta\beta^{h,error}$ on the resulting $\beta^{h,end}$ and on the resulting MSE can be calculated in a similar way as done in section 6.3.

Assuming that the quantization error $\Delta\beta^{h,error}(q)$ is proportional to the quantization step q [5], it follows that the left had side of (7.19) is equivalent to

$$\Delta\beta^{h,error}(q) \approx q \left. \frac{\partial \displaystyle\int \frac{P(D-Y)}{P(correct)}\, Q\varepsilon\left[(D-Y) f'(f^{-1}(Y)) f^{-1}(Y), q\right] dY}{\partial q} \right|_{q \ll q_{crit}} \tag{7.20}$$

Substituting (7.20) in (7.19) gives the following simple equation for the maximum tolerable q:

$$q \leq \frac{\eta}{10\beta_{ideal}^{h,end}} \left(\frac{\displaystyle\int_{Lin.Sep.} \frac{P(D-Y)}{P(correct)} (D-Y) f'(f^{-1}(Y)) f^{-1}(Y)\, dY}{\left. \dfrac{\partial \displaystyle\int \frac{P(D-Y)}{P(correct)}\, Q\varepsilon\left[(D-Y) f'(f^{-1}(Y)) f^{-1}(Y), q\right] dY}{\partial q} \right|_{q \ll q_{crit}}} \right) \tag{7.21}$$

For the distribution of the errors, $\mathcal{P}(D-Y)$, one may use one of the two estimations presented in Appendix 6.2, which are sufficiently accurate.

Numerical illustration

Calculations show that both the numerator and the denominator in the expression between brackets in (7.21) are approximately independent of $\mathcal{P}(D-Y)$ and q. For a sigmoid non-linear function with unit maximum derivative and a double threshold at 0.05 and 0.95 of the maximum response of the neuron, calculated values for the denominator are typically in the range [0.01, 0.02], while the numerator is typically approximately 0.05[1].

Therefore, the maximum tolerable quantization step q is (for this relatively simple training set) worst case given by

$$q \leq \frac{0.2\,\eta}{10\beta_{ideal}^{h,end}}, \quad \left(f(x) = \frac{1}{1+exp(-4x)}, \quad D \in \{0.05, 0.95\}\right). \tag{7.22}$$

[1] This last value can easily be explained as follows: the magnitude of the quantization errors is $0.5 \cdot q$ while the quantization errors cancel in first order. First order cancellation of errors decreases errors typically by one order of magnitude giving the factor 0.05.

7.4.2 A less simple case

In sub-section 7.4.1, an estimation for the maximum tolerable quantization step q in order not to affect the learning behavior significantly is derived for a simple training set. In that sub-section it was assumed that $\alpha^{ATT}=0$ and that the $\underline{B}^h$ vector has one non-zero element. In this sub-section, it is still assumed that the $\underline{B}^h$ vector has only one non-zero element, but α^{ATT} can have any value.

In this case, (7.18) can be rewritten into

$$\Delta\beta^{h,error}(q) \approx \int \wp(D-Y)\frac{Q\varepsilon\left[\eta(D-f(\underline{W}\cdot\underline{U}))f'(\underline{W}\cdot\underline{U})(\alpha^{ATT}+\alpha^{h,\nu}),q\right]}{P(correct)}\,d\alpha^{h,\nu}. \qquad (7.23)$$

Figure 7.8 shows the calculated (using (7.23)) $\Delta\beta^{h,error}$ as a function of the quantization step q. The calculations were done for the same non-linearly separable training set as used for figure 7.5. For comparison reasons, the x-axis has been taken identical to the x-axis in figure 7.5, and the two curves in figure 7.5 have been included (dotted) in figure 7.8.

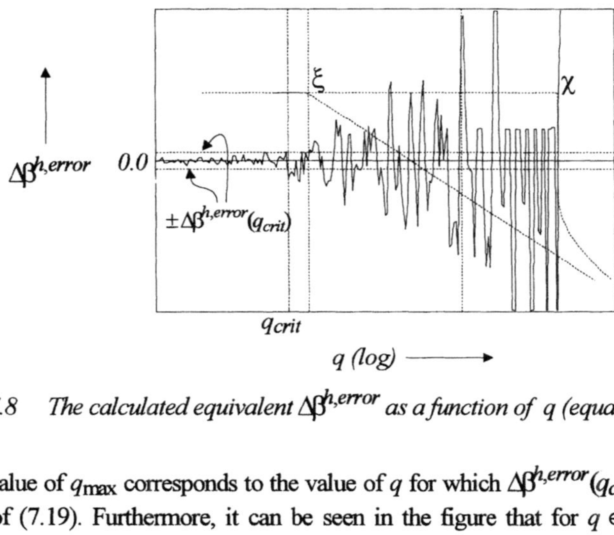

Fig. 7.8 The calculated equivalent $\Delta\beta^{h,error}$ as a function of q (equation (7.23))

The value of q_{max} corresponds to the value of q for which $\Delta\beta^{h,error}(q_{crit})$ equals the right hand side of (7.19). Furthermore, it can be seen in the figure that for $q \in [\xi, \chi]$, the $\Delta\beta^{h,error}$ is heavily dependent on the exact value of q. This dependency explains the fact that the simulated $\beta^{h,end}$ in figure 7.5 "oscillates" around $\beta^{h,end}_{ideal}$ in this particular range.

It follows that although the absolute maximum quantization step q allowed in order to reach a specific $\beta^{h,end}$ increases with $|\alpha^{ATT}|$, a worst-case estimation of the maximum tolerable quantization step in order to have a negligibly small effect of the quantization is given by (7.21) and (7.22).

7.4.3 A general case

Equations (7.21) and (7.22) give a maximum tolerable quantization step size q in order to have a negligibly small effect of quantization on training, for the two cases in sections 7.4.1 and 7.4.2. For larger quantization steps q than indicated by (7.21) and (7.22), the neuron may learn the training set, but the eventual performance is heavily dependent on both the training set and on the quantization step.

In this sub-section, a worst-case estimation for a general training set is presented: it is assumed that the $\underline{B}^h$ vector has $(N_{in}-1)$ non-zero elements. In this case, the quantization of the weight adaptations will affect the adaptation of all elements of $\beta^h\underline{B}^h$ differently. With the conclusions of chapter 6 and section 7.3.3, it follows that worst-case all elements of the $\underline{B}^h$ are equal:

$$\underline{B}^h_{wc} = [a\,, a\,, a, \ldots, a\,, 0]^T$$

$$with \quad a = (N_{in}-1)^{-1/2}$$

For this case, the *worst-case* the quantization step q must satisfy

$$\left|\Delta\beta^{h,error,wc}(q)\right| \leq \frac{\eta}{10\,\beta^{h,end}_{ideal}} \int\limits_{Lin.Sep.} \mathcal{P}(D-Y)\frac{(D-Y)f'\left(f^{-1}(Y)\right)f^{-1}(Y)}{P(correct)}\,d(D-Y). \qquad (7.24)$$

$$\text{with } \Delta\beta^{h,error,wc}(q) \approx \sqrt{N_{in}-1}\int \mathcal{P}(D-Y)\frac{Q\varepsilon\left[\eta(D-f(\underline{W}\cdot\underline{U}))f'(\underline{W}\cdot\underline{U})\frac{\alpha^{h,v}}{\sqrt{N_{in}-1}}, q\right]}{P(correct)}\,d\alpha^{h,v}$$

The factor $\sqrt{N_{in}-1}$ in (7.24) results from equal quantization errors on all $(N_{in}-1)$ non-zero elements of $\beta^h\underline{B}^h$. Assuming that the average quantization error $\Delta\beta^{h,error}(q)$ is proportional to the quantization step q [5], a worst case estimation for the tolerable quantization step size q can be given

$$q \leq \frac{\eta}{10\beta^{h,end}_{ideal}\sqrt{N_{in}-1}}\left(\frac{\displaystyle\int\limits_{Lin.Sep.}\frac{P(D-Y)}{P(correct)}\left(D-Y\right)f'(f^{-1}(Y))f^{-1}(Y)\,dY}{\left.\dfrac{\partial\displaystyle\int\frac{P(D-Y)}{P(correct)}Q\varepsilon\left[(D-Y)f'(f^{-1}(Y))f^{-1}(Y), q\right]dY}{\partial q}\right|_{q\ll q_{crit}}}\right)$$

Identical to (7.21), both the numerator and the denominator in the expression between brackets are approximately independent of $\mathcal{P}(D-Y)$ and q.

Numerical illustration
For a sigmoid non-linear function with unit maximum derivative and a double threshold at 0.05 and 0.95 of the maximum response of the neuron, calculated values for the denominator are typically in the range [0.01, 0.02], while the numerator is typically approximately 0.05. The maximum tolerable quantization step q is worst case

$$q \leq \frac{0.2\,\eta}{10\beta_{ideal}^{h,end}\sqrt{N_{in}-1}}, \quad \left(f(x)=\frac{1}{1+exp(-4x)},\ D \in \{0.05, 0.95\}\right). \tag{7.25}$$

Because the maximum usable learning rate η decreases linearly with N_{in} (see chapter 3) in order to ensure stability during learning, it follows that the worst case maximum quantization step scales proportional to $(N_{in})^{-1\frac{1}{2}}$. Note that with the estimations in section 7.5, estimating the required digital resolution during training is straight-forward.

7.5 Estimation of absolute minimum weight-storage C size

The minimum required quantization step size as derived in this chapter can also be used to derive the *absolute* minimum value for the size of the weight storage capacitor in analog on-chip learning neural networks (see also chapter 6). By noting that the charge on a capacitor has discrete values (multiples of the charge of one electron), a worst-case estimation for the absolute minimum size for the weight storage capacitor is given by

$$C^{weight} \geq \frac{e\,\hat{W}}{\hat{V}^{weight}\,q}$$

where

e	is the charge of one electron
$[-\hat{W}, \hat{W}]$	is the implemented weight range
$[-\hat{V}^{weight}, \hat{V}^{weight}]$	is the corresponding weight voltage range
q	is given by (7.21)

Note that this estimation assumes that the charge loaded onto the weight storage capacitor can perfectly be controlled. For the same values as used in the worst-case estimation of the minimum required size of the weight storage capacitor when using non-ideal weight adaptation circuitry (section 6.7), $\hat{V}^{weight}=1.5$, $\hat{W}=10$, $\eta=0.1$ and $N_{in}=5$ it follows that

$$C^{weight} \geq 0.01\,pF$$

Which is about four orders of magnitude smaller than for the case that the weight adaptation circuit causes a systematic charge injection of 100 *electrons* at each weight adaptation, see section 6.7.

7.6 Conclusions

This section presents mathematical estimations of precision requirements for single-layer feed-forward neural nets, both in analog and in digital implementations. It is shown that for a specific non-linearity and specific threshold values, the worst-case precision depends only on the learning rate, the eventual norm of the weight vector and on the dimension of the input space. For precisions lower than those indicated by the worst-case estimations in this chapter, the neural network may learn the training set, but the eventual performance will be heavily dependent on both the precision and on the training set.

7.7 References

[1] K.Asanovic and N.Morgan, "Experimental Determination of Precision Requirements for Back-Propagation Training of Artificial Neural Networks", In Proceedings of the 2nd International Conference on Microelectronics for Neural Networks, Munich: Kyrill & Method Verlag, pp. 9-16, 1991

[2] R.C.Frye, E.A.Rietman and C.C.Wong, "Back-Propagation Learning and Nonidealities in Analog Neural Network Hardware", *IEEE transactions on Neural Networks*, vol.2, pp. 110-117, 1991

[3] R.Gregorian and G.C.Themes, *Analog MOS Integrated Circuits for Signal Processing*, New York: Wiley, 1986

[4] P.W.Hollis, J.S.Harper and J.J.Paulos, "The Effects of Precision Constraints in a Backpropagation Learning Network", *Neural Computation*, vol. 2, pp. 363-373, 1990

[5] D.E.Rumelhart, G.E.Hinton and R.J.Williams, "Learning internal representations by error propagation", *Parallel Distributed Processing: Exploration in the microstructure of cognition* (Vol. 1), Cambridge, MA.:MIT Press, 1986

[6] S.Sakaue, T.Kohda, H.Yamamoto, S.Maruno and Y.Shimeki, "Reduction of Required Precision for Back-Propagation Applied to Pattern Recognition", *IEEE transactions on Neural Networks*, vol. 2, pp. 270-275, 1993

[7] Y.Xie and M.A.Jabri, "Analysis of Effects of Quantization in Multilayer Neural Networks using Statistical Model", *Electronics Letters*, vol. 27, pp. 1196-1198, 1991

[8] H.Yoo and R.L.Pimmel, "The effect of weight precision and range on neural network classifier performance", Neurocomputing, vol. 6, no. 5-6, pp. 541-549, 1994

Chapter 8

Learning Behavior and Temporary Minima of Two-Layer Neural Networks

8.1 Introduction

Considerable research has been done in the field of multi-layer neural networks with respect to feed-forward network abilities [1],[2],[10],[14]. The dynamic behavior of a single-layer neural network has been well analyzed [15],[21],[chapters 3 to 7]. However, the learning behavior (or dynamics) during training of a multi-layer neural network is generally derived by simulations because mathematical analyses of dynamics of non-linear systems such as a multi-layer neural network are very complicated. Only very few mathematical analyses of the dynamics of a feed-forward multi-layer network have been published [5]. Although mathematical analyses may require drastic simplifications in order to derive easy to read equations, they contribute much more to the understanding of the fundamental principles involved than only simulations do.

In this chapter, a major problem during multi-layer feed-forward neural network training is analyzed using the VDM. During multi-layer perceptron learning, undesired minima may be encountered. In undesired minima, the performance improvement of the classification drops to very low levels or even approaches zero while the network has not (yet) arrived at the optimal state for classification of the training set. Two types of undesired minima can be distinguished, local minima [8] and temporary minima [16]. The origin of temporary minima are analyzed in this chapter.

Temporary minima

As mentioned in section 8.1, two types of minima can be distinguished during multi-layer neural network training: the local minimum and the temporary minimum. In this section, these two types of minima will be discussed briefly.

- The performance improvement of the classification drops to zero if the network is getting stuck in a *local minimum* in the energy landscape. The local minimum may be abandoned by adding noise or by using sophisticated convergence algorithms. Possible modifications include simulated annealing [17],[20] and applying another energy function [6]. Note that a globals minimum in the energy landscape is merely a special type of local minimum.
- *Temporary minima* [16] are fundamentally different from local minima as the performance improvement in this type of minimum drops to a very low but non-zero level because an approximately flat region in the energy landscape is encountered. Without modifications to the training set or learning algorithm, the network may escape from this type of 'minimum' but performance improvement in these temporary minima is very low because of the very low gradient in the energy landscape. In the mean squared error (MSE) versus training time curve, a temporary minimum can be recognized as a phase in which the MSE is virtually constant for a long training time after initial training. After a generally long training time, the approximately flat part in the energy landscape is abandoned, resulting in a significant and 'sudden' drop in the MSE curve [16],[24]. An illustration of a *MSE-training time* curve with two temporary minima is shown in figure 8.1.

As temporary minima slow down learning significantly, minimizing the training time that is spent in this type of undesired minima speeds up learning considerably. In this chapter, the fundamental process behind the temporary minima is derived mathematically under condition of small initial weights. This condition will generally be satisfied in neural network training if no a priori knowledge about the training set is included in initial weights. The applicability of the presented analysis if the condition of small initial weights is not fulfilled is also discussed.

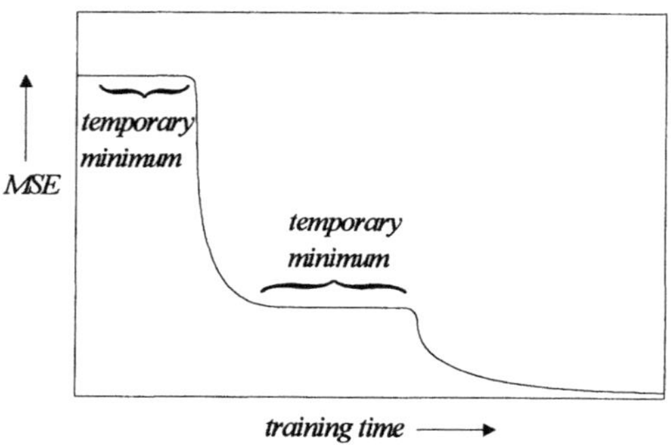

Fig. 8.1 A learning curve (MSE as a function of the training time) with two temporary minima

Outline of the chapter
The outline of this chapter is as follows. A short summary of the notation used for multi-layer neural networks, of the backpropagation algorithm, and of the VDM applied to multi-layer neural nets is given in section 8.2. Also a short *preview of the analysis* is provided in this section.
The initial training and assumptions used for the analyses in this chapter are discussed in the first part of section 8.3. The remainder of section 8.4 presents an introduction into the analyses of two (strongly related) mechanisms, called the "*rotation-based*" and "*translation-based*" mechanisms, that lead towards the encounter of temporary minima. The rotation-based mechanism is analyzed in section 8.4; an extensive discussion of this mechanism is included in this section. An illustrative example of this mechanism is presented in section 8.5.
The analysis and discussion of the translation-based mechanism are presented in section 8.6. As the analysis of this mechanism is strongly related to that of the rotation-based mechanism, the analysis and discussions are kept short. In section 8.7, the XOR-problem is selected to briefly illustrate the conclusions of section 8.6.
The analyses in sections 8.3 to 8.7 will be done for a two-layer network with two neurons in the first layer. An extension towards neural networks with an arbitrary number of neurons in the first layer is given in section 8.8.
Section 8.9 summarizes the conclusions.

8.2 A summary

In this section, a short summary is provided for a number of topics. Firstly, the notation used for multi-layer neural networks and the type of multi-layer feed-forward neural network analyzed in this chapter are discussed. After this, a short summary of the backpropagation algorithm is given. The analyses in this chapter are done using the VDM, which was introduced in chapter 2. Last of all, the last part of this section provides a short *preview* of the analyses in this chapter.

8.2.1 The network and the notation

In this chapter, the origin of temporary minima during training of two-layer neural networks is analyzed. In the analyses, it will be assumed that there be only one output signal of the network: the analyses assume a *single-output two-layer feed-forward neural network* (see also section 1.2.1). A general single-output two-layer feed-forward network is shown in figure 8.2.

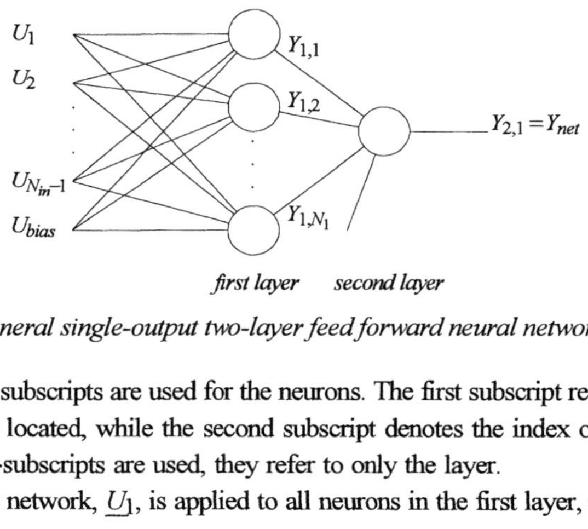

Fig. 8.2 General single-output two-layer feed forward neural network

In the notation, double subscripts are used for the neurons. The first subscript refers to the layer in which the neuron is located, while the second subscript denotes the index of the neuron in the layer. When single-subscripts are used, they refer to only the layer.

The input vector of the network, $\underline{U_1}$, is applied to all neurons in the first layer, and is given by

$$\underline{U_1} = \left[U_1,\, U_2,\, \ldots, U_{N_{in}-1},\, U_{bias}\right]^T. \tag{8.1}$$

The input vector of the neuron in the second layer consists of all output signals of the neurons in the first layer, $Y_{layer,index}$, and of a (constant) bias input signal:

$$\underline{U_2} = \left[Y_{1,1},\, Y_{1,2},\, \ldots, Y_{1,N_1},\, U_{bias}\right]^T. \tag{8.2}$$

The output signal of the single neuron in the second layer is the output signal of the neural network. In formula:

$$Y_{2,1} = Y_{net} \tag{8.3}$$

8.2.2 Back-propagation rule

In the previous chapters, the back-propagation rule was used to train single-layer neural networks. In this chapter, the analyses assume that the (double-threshold variant of the) back-propagation rule is used to adapt the weights of the two-layer neural network. Some basics of the back-propagation rule are briefly summarized in this sub-section.

With the back-propagation rule, first a mean squared error cost function E defined. In formula, the cost function is

$$E = \frac{1}{2P}\sum_{p=1}^{P}\left(D^p - Y^p\right)^2 \tag{8.5}$$

where

p	denotes the index of the training example
P	is the number of training examples in the training set
Y^p	is the network response on the pth training example
D^p	is the desired (or target) response for the pth example

For pattern classification, the desired responses D^p are either "low" or "high". For the training, we use the well known *double-threshold* modification of the back-propagation algorithm [8], which states that there be no adaptation for examples that satisfy

$$Y^p \geq D^p\Big\backslash_{D^p=D^{high}} \quad \text{and} \quad Y^p \leq D^p\Big\backslash_{D^p=D^{low}} \tag{8.6}$$

During training, the cost-function E is minimized using a gradient descent method. For the adaptation of the weight vector of the neurons in the second layer, it follows that

$$\Delta \underline{W}^p_{2,1} = -\eta\frac{\partial E^p}{\partial \underline{W}_{2,1}} = \eta\,(D^p - Y^p)\,f'(\underline{W}_{2,1}\cdot\underline{U}^p_2)\,\underline{U}^p_2. \tag{8.7}$$

For the adaptation of the weight vector of a neuron in the first layer, one requires the chain-rule. It can be derived that the adaptation is given by:

$$\Delta \underline{W}^p_{1,n} = -\eta\frac{\partial E^p}{\partial \underline{W}_{1,n}} = \eta(D^p - Y^p)f'(\underline{W}_{2,1}\cdot\underline{U}^p_2)\,(\underline{W}_{2,1})_n\, f'(\underline{W}_{1,n}\cdot\underline{U}^p_1)\,\underline{U}^p_1. \tag{8.8}$$

where

$(\underline{W}_{2,1})_n$	denotes the nth element in the weight vector of the neuron in the second layer

8.2.3 Vector decomposition

For the analysis of the dynamics during training of two-layer neural networks, we use the VDM which was introduced in chapter 2. With the VDM, the weight vectors and input vectors of all neurons in the network are decomposed into three vector components which are related to so-called *attractor hyperplanes*. This section briefly summarizes the concept of attractor hyperplanes and presents the notation of the weight- and input vectors in multi-neuron networks.

Attractor hyperplanes

A multi-layer neural network is known to be a universal approximator [9]. Consequently, the weight vectors of the neurons in the network converge during training towards specific attractors in weight space [5],[18]. The actual attractors may depend on the initial weights. For reasons of conceptual simplicity, we assume the initial weights to be "very small". The condition for "very small" initial weights will be defined in section 8.4.

The attractor in weight space for a specific neuron will be denoted as the *weight vector attractor*, $\underline{W}_{kj}^{ATT}$. Because of the different phases that will be shown to be surpassed during training in this chapter, the weight vector attractor is generally different in different phases. The hyperplane corresponding to the weight vector attractor will be denoted as the *attractor hyperplane* and will henceforth generally be different in different phases during training.

Vector decomposition

With the VDM we decompose the weight vector and the input vector of every neuron in the neural network into three orthogonal vector components related to its attractor hyperplane, see chapter 2. One of the vector components is related to only the bias input of the neurons, one component is perpendicular to the attractor hyperplane and the last vector component is in parallel to the attractor hyperplane of the corresponding neuron.

For the input vector and the weight vector of the jth neuron in the kth layer we obtain:

$$\underline{U}_{kj} = \alpha_{kj}^{h}\,\underline{B}_{kj}^{h} + \alpha_{kj}^{bias}\,\underline{B}_{kj}^{bias} + \underline{U}_{kj}^{F} \tag{8.7}$$

and

$$\underline{W}_{kj} = \beta_{kj}^{h}\,\underline{B}_{kj}^{h} + \beta_{kj}^{bias}\,\underline{B}_{kj}^{bias} + \underline{W}_{kj}^{\varepsilon}\,. \tag{8.8}$$

The $\underline{B}_{kj}^{h}$ vectors are unit vectors perpendicular to the attractor hyperplane of the jth neuron in the kth layer and the $\underline{B}_{kj}^{bias}$ vector is a unit vector which is related to only the bias input signal of the neuron. The vector components that are in parallel to the attractor hyperplane are represented by $\underline{U}_{kj}^{F}$ and $\underline{W}_{kj}^{\varepsilon}$ and need not to be decomposed into unit vectors and scalar multiplication factors for the analyses in this chapter.

As mentioned in section 2.4.3, the α_{kj}^{h} and β_{kj}^{h} may be associated with the concepts of relevant information and correct knowledge respectively. Similarly, $|\underline{U}_{kj}^{F}|$ and $|\underline{W}_{kj}^{\varepsilon}|$ may be associated with irrelevant information and with incorrect knowledge respectively.

8.2.4 Preview of the analyses

During training of two-layer neural networks, temporary minima may be encountered. In the encounter of temporary minima, several distinguishable phases will be surpassed. In each of these subsequent training phases, the attractors may be different from the previous attractor positions. This will become clear along with the analyses that follow in sections 8.4 to 8.9. At this point, we like to ask the reader to keep up with the analysis, starting with the assumption of “very small” initial weights. The analysis will take the following line.

The first phase: initial training

In the first phase of training using “very small” initial weights, the neural network can be linearized because all neurons are activated in the approximately linear middle region. In this phase, redundancy is built up in the first layer; in others words, the attractors in weight space for all neurons in the first layer are identical. After this phase, the first layer is approximately reducible to one neuron [22] and the weight vectors of the neurons in the first layer are approximately identical.

The second phase: encounter of a temporary minimum

In the next phase, the network can not be linearized any more and the network non-linearities must be included in the analysis. In this phase, the cluster of redundant neurons in the first layer will first continue building up redundancy and may then start to abolish this redundancy partially by starting to 'break' into two clusters of neurons. During this phase, the attractors in weight space are still identical for all neurons in the first layer. It will be shown in the following sections of this chapter that in this phase the neural network "sticks" in a temporary minimum.

The third phase: escape from a temporary minimum

A third phase continues the "breaking" of the cluster into two smaller clusters. If the cluster is effectively 'broken' into two clusters, the weight vector of the neurons in each cluster now converge towards two significantly different attractors in weight space. It will be shown in this chapter that this "breaking" of clustered hyperplanes results in the "sudden" escape of temporary minima.

Encountering multiple temporary minima

The second and third phase may iteratively be repeated a number of times, which gives rise to the subsequent encounter and leaving of multiple temporary minima. An illustration of the subsequent surpasing of all phases during training is presented in the following figure.

An illustration

Figure 8.3 presents an illustration of the three phases to be sequentially surpasses during training of a two-layer neural network. Note that the learning curve (*MSE* versus training time) shown is identical to the one in figure 8.1. In the figure, the three phases discussed earlier are marked by the numbers 1 to 3.

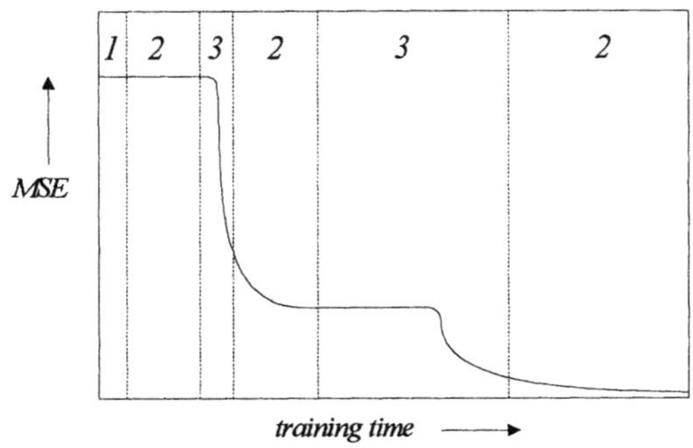

Fig. 8.3 The three phases (1 to 3) surpassed during training, and their correspondence with temporary minima

In sections 8.3 to 8.8, the mechanims that have been described briefly in this section are analyzed mathematically using the VDM. It follows that two mechanisms can lead towards the encounter and subsequent escape of temporary minima; the mechanisms have been denoted as the "*rotation based breaking*" mechanism and the "*translation based breaking*" mechanism. An introduction into the analysis and a discussion of the two mechanisms is presented in the following section, 8.3.

The analyses of the two mechanisms and illustrative examples are presented in sections 8.4 to 8.7 for cases in which one temporary minima is encountered during training. The encounter of multiple temporary minima is discussed in section 8.8.

8.3 Analysis of temporary minima: introduction

The vector decomposition method will now be used to analyse the dynamic behavior of a learning neural network when, after initial training, a temporary minimum is encountered. In a temporary minimum, the energy landscape is approximately flat and hence the MSE is approximately constant for a relatively long training time. It is observed that after a long training time, the mean squared error 'suddenly' drops to a significantly lower level, see figure 8.1.
Encountering a temporary minimum slows down the learning process significantly. Consequently, reducing the time during which the network 'sticks' in this type of minima would speed up learning significantly. In this section, first a description of the learning behavior of two-layer neural networks in the very beginning of training is presented. For this description, it is assumed that the initial weights are small. The condition for small initial weights is usually satisfied in neural network training when no a priori knowledge about the training set is included in initial weights. After this description, a short introduction of two related mechanisms (rotation-based and translation-based) that lead towards the encounter of temporary minima is presented. The rotation-based mechanism is analyzed in detail in section 8.4, and the translation-based mechanism is analyzed in 8.6.

Assumptions
In order to obtain easy-to-read equations, we make some assumptions concerning the weight vectors of the neurons in the network and concerning the number of neurons in the network. These assumptions do not degenerate the generality of this approach.
Firstly, it is assumed that the *number of neurons in the first layer* of the network is *two*. In the analysis, any of these two neurons can be replaced by a cluster of (redundant) neurons without modifications to the analysis. Furthermore, it will be shown by induction in section 8.8 that the analyses presented in sections 8.3 to 8.6 hold for any single-output two layer network with an arbitrary number of neurons in the first layer.

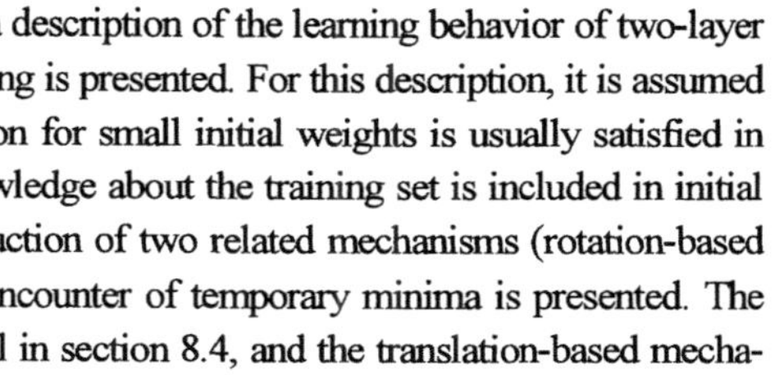

Fig. 8.4 It is first assumed that there be only two neurons in the first layer; section 8.8 presents an extension

A second assumption is that the initial weights of the neurons in the network be small enough in order to ensure that all neurons are activated in the approximately linear middle region of the transfer function during the beginning of training. Initial weights satisfying this assumption are *"very small"* initial weights. As long as the weights are very small, the neural network can be linearized. Initial weights that are not very small are also discussed in section 8.8.

8.3.1 Initial training: a linearized network

In this section, a discussion of initial training of a two-layer neural network with "very small" initial weights is presented. It will be shown that during initial training, the weight vector of first layer neurons adapt closely identical and weights in the second layer's neuron adapt closely identical. The results of this section serve as a starting point for the analyses of the encounter of temporary minima in sections 8.4 and 8.6.

All non-bias elements in $\underline{W}_{2,1}$ adapt (closely) identical

At the start of training, the weights of the neurons in the first layer are very small. The weighted input of neurons in the first layer will then be close to zero for any training example. Consequently, the response of any neuron in the first layer on any training example is approximately $f(0)$. In formula:

$$\{\forall \underline{U}^p \mid \underline{W}_{1,j} \cdot \underline{U}_1{}^p \approx 0\} \Rightarrow \{\forall \underline{U}^p \mid f\left(\underline{W}_{1,j} \cdot \underline{U}_1{}^p\right) \approx f(0)\}$$

As already stated in (8.7), the adaptation of a weight of the neuron in the second layer for any single training example is given by

$$\Delta(\underline{W}_{2,1})_n = \eta\,(D-Y)\,f'(\underline{W}_{2,1} \cdot \underline{U}_2)\,(\underline{U}_2)_n.$$

The first three factors on the right hand side of this equation are shared for all weights of the neuron. Only the last factor, denoting the nth element of the $\underline{U}_2$ vector is different for the adaptation of the elements in $\underline{W}_{2,1}$. In the previous paragraph, it has been shown that for very small initial weights, the responses of the neurons in the first layer are closely identical. The weights in $\underline{W}_{2,1}$ connected to neurons in the first layer will therefore adapt identically on any training example in the very beginning of training.

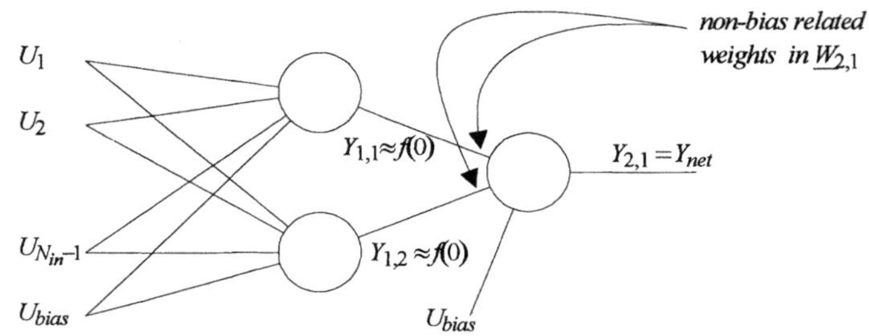

Fig. 8.5 The non-bias weight in $\underline{W}_{2,1}$ adapt closely identical in the beginning of training

The weight vectors of all first layer neurons adapt (closely) identical
The adaptation of the weight vectors of the neurons in the first layer is given by

$$\Delta \underline{W}_{1,j} = \eta (D-Y) f'(\underline{W}_{2,1} \cdot \underline{U}_2)(\underline{W}_{2,1})_j f'(\underline{W}_{1,j} \cdot \underline{U}_1)\, \underline{U}_1.$$

The first three factors on the right hand side on this equation are again identical for all first layer neurons. Furthermore, as the elements in $\underline{W}_{2,1}$ corresponding to neurons in the first layer adapt identically on any training example, the weights $(\underline{W}_{2,1})_j$ are closely identical (the forth factor in the expression). The fifth factor on the right hand side of the equation, $f'(\underline{W}_{1,j} \cdot \underline{U}_1)$, is identical for all neurons assuming either very small weights or almost identical weight vectors $\underline{W}_{1,j}$. The input vector for the first layer $\underline{U}_1$ is finally fed to all neurons in the first layer and is hence identical for all neurons in the first layer.
Therefore, it can be concluded that in the beginning of training, the weight vectors $\underline{W}_{1,j}$ adapt by approximation identically on any training example and that furthermore the associated weights $(\underline{W}_{2,1})_j$ adapt closely identical[1]. In formula:

$$(\underline{W}_{2,1})_i \approx (\underline{W}_{2,1})_j$$

For global training, the weights are updated after presentation of the total training set. The weight updates for $(\underline{W}_{2,1})_j$ and the weight vector adaptations $\Delta \underline{W}_{1,j}$ are approximately equal for any training example. Consequently, in case of global learning, the summed weight updates over the total training set are also closely identical for the neurons in the first layer and for the weights in $\underline{W}_{2,1}$ associated with the output of these neurons.

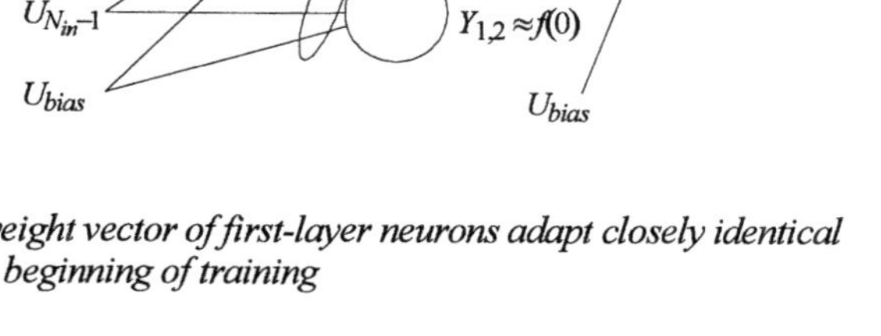

Fig. 8.6 The weight vector of first-layer neurons adapt closely identical in the beginning of training

[1] Note that there is a possibility that the weight vectors of the neurons in the first layer adapt in *opposite* directions, which introduces opposite signs for the associated weights of the neuron in the second layer. This situation will be discussed later on in this chapter.

An illustration of the process of closely identical weight adaptation in the beginning of training is presented in figure 8.8 in section 8.5. The figure presents the dynamics of weights in weight space for the training set of figure 8.7, as simulated by a neural network simulation program. In the beginning of training, i.e. for $(\underline{W}_{k,j})_n$ small, the adaptation of $\underline{W}_{1,j}^{r}$ appears to proceed almost identically. The same holds for the corresponding weights of the neuron in the second layer, $(\underline{W}_{2,1})_j$.

Notation with the VDM

In the very beginning of the training and under condition of very small initial weights, the weight vectors of the neurons in the first layer adapt closely identically. The corresponding hyperplanes of the neurons in the first layer move therefore towards the same position in input space. Moving towards identical positions in input space of the hyperplanes can be interpreted as coinciding attractor hyperplanes for the two neurons in the first layer of the network:

$$\underline{B}_{1,1}^{h} = \underline{B}_{1,2}^{h} \equiv \underline{B}_1^{h} \tag{8.9}$$

As the weight vectors of the two neurons in the first layer adapt almost identical and with (8.9), the $\beta_{1,j}^{h}$ of the two neurons in the first layer adapt to approximately identical values as long as the network can be linearized. In the analysis, it is assumed for reasons of simplicity that

$$\beta_{1,1}^{h} = \beta_{1,2}^{h} \equiv \beta_1^{h}. \tag{8.10}$$

The weight vectors of the neurons in the first layer will, under condition of proper initialization converge towards the same weight vector attractor. Based on the analysis of the dynamics of the neural network using very small initial weights, i.e. weights that allow linearization of the network in early training, we posit the following theorem:

Theorem 8.1: On condition of very small initial weights, the hyperplanes of the neurons in the first layer of a single-output two-layer neural network will, in the first part of training, move towards identical positions in input space.

The weight vectors will however not be *exactly* identical; they will generally have different $\underline{W}_{k,j}^{\varepsilon}$ and $\beta_{k,j}^{bias}$ vector components. Due to the fact that the neural network can be linearized, the classification made by the network in the very beginning of training corresponds to the classification made by the average position of the hyperplanes of the neurons in the first layer. Because the weight vectors of both neurons in the first layer attract to their weight vector attractors $\underline{W}_1^{ATT}$,

$$\frac{1}{2}\left(\underline{W}_{1,1} + \underline{W}_{1,2}\right) = \underline{W}_{1,1}^{ATT} = \underline{W}_{1,2}^{ATT} = \underline{W}_1^{ATT}\ . \tag{8.11}$$

With the vector decomposition method, the $\underline{W}_1^{ATT}$ is decomposed as

$$\underline{W}_1^{ATT} = \beta_1^h \underline{B}_1^h + \beta_{ATT}^{bias} \underline{B}_1^{bias}. \tag{8.12}$$

It follows from (8.11) and (8.12) that as long as the network can be linearized, (8.13) will be satisfied by good approximation:

$$\begin{cases} \underline{W}_{1,1}^{\varepsilon} = -\underline{W}_{1,2}^{\varepsilon} \\ \beta_{1,1}^{bias} + \beta_{1,2}^{bias} = \beta_{ATT}^{bias} \end{cases} \tag{8.13}$$

For simplicity of presentation, both a vector $\underline{\Delta W}^{\varepsilon}$ denoting the difference of the $\underline{W}_{1,j}^{\varepsilon}$ vectors and a differential norm of the $\underline{B}_{1,j}^{bias}$ components are introduced. These new components satisfy:

$$\begin{cases} \underline{\Delta W}^{\varepsilon} = \underline{W}_{1,1}^{\varepsilon} - \underline{W}_{1,2}^{\varepsilon} \\ \Delta\beta_1^{bias} = \beta_{1,1}^{bias} - \beta_{1,2}^{bias} \end{cases}. \tag{8.14}$$

It follows that when starting with "very small" initial weights that allow linearization of the network in the beginning of training, the weight vectors of the two neurons in the first layer adapt to

$$\underline{W}_{1,1} = \beta_1^h \underline{B}_1^h + \left(\beta_1^{bias} + \tfrac{1}{2}\Delta\beta_1^{bias}\right)\underline{B}_1^{bias} + \tfrac{1}{2}\underline{\Delta W}^{\varepsilon} \tag{8.15a}$$

$$\underline{W}_{1,2} = \beta_1^h \underline{B}_1^h + \left(\beta_1^{bias} - \tfrac{1}{2}\Delta\beta_1^{bias}\right)\underline{B}_1^{bias} - \tfrac{1}{2}\underline{\Delta W}^{\varepsilon}. \tag{8.15b}$$

in the beginning of training.

It was also shown that the weights of the neuron in the second layer associated with neurons in the first layer, $(\underline{W}_{2,1})_n$, adapt (by approximation) identically in the beginning of training. It will first be assumed that the signs of these weights are equal; a discussion of a more general case will be given later on in this chapter.

$$(\underline{W}_{2,1})_n = w_{2,1} \quad n \in \{1, ..., N_1\} \qquad \text{with } N_1 = 2. \tag{8.16}$$

With (8.16), the $\underline{B}_{2,1}^h$ vector is

$$\underline{B}_{2,1}^h = \begin{bmatrix} \sqrt{1/N_1} \\ \sqrt{1/N_1} \\ 0 \end{bmatrix} = \begin{bmatrix} \sqrt{0.5} \\ \sqrt{0.5} \\ 0 \end{bmatrix} \tag{8.17}$$

8.3.2 Continued training: including network non-linearities

The analysis of the learning behavior of two-layer networks as presented assumes that the weight vectors of the neurons in the first layer be close enough to satisfy condition (8.18).

$$f(\underline{W}_{1,1}\cdot\underline{U}_1)+f(\underline{W}_{1,2}\cdot\underline{U}_1)\approx 2f(\underline{W}_1^{ATT}\cdot\underline{U}_1) \tag{8.18}$$

With (8.18), the response of the two layer neural network on an input vector $\underline{U}_1$ is

$$Y(\underline{U}_1)=f\left((\underline{W}_{2,1})_1 f(\underline{W}_{1,1}\cdot\underline{U}_1)+(\underline{W}_{2,1})_2 f(\underline{W}_{1,2}\cdot\underline{U}_1)+\alpha_{2,1}^{bias}\,\beta_{2,1}^{bias}\right).$$

With (8.15) to (8.18), this equation can be approximated by

$$Y(\underline{U}_1)=f\left(\sqrt{2}\,\beta_{2,1}^{h}f(\underline{W}_1^{ATT}\cdot\underline{U}_1)+\alpha_{2,1}^{bias}\,\beta_{2,1}^{bias}\right). \tag{8.19}$$

Note that the factor $\sqrt{2}$ in the first term of the argument springs from $N_1\sqrt{1/N_1}=\sqrt{2}$ for a neural net with two neurons in the first layer, i.e. $N_1=2$.

Condition (8.18) includes (8.11), but is valid in a larger range: condition (8.18) is not only satisfied for small weights, i.e. weights that allow linearization, but is *also satisfied* when the weight vectors $\underline{W}_{1,1}$ and $\underline{W}_{1,2}$ are close. This includes that for large weights, (8.18) is satisfied if the angle between $\underline{W}_{1,1}$ and $\underline{W}_{1,2}$ is small and at the same time the hypererplanes are close for examples in the training set. As soon as (8.18) is not satisfied any more, the two neurons in the first layer make a significantly different classification. In this case, the analysis of the learning behavior can be continued after making a new vector decomposition for both neurons in the first layer. In sections 8.5, 8.7 and 8.9, making a new decomposition will be explained in more detail.

Declustering mechanisms: an introduction

It has been shown that during initial training, the hyperplanes corresponding to the neurons in the first layer attract to the same position in input space. When including the network non-linearities, the so-formed cluster of hyperplanes may "break". Two mechanism that result in "breaking" of the clustered hyperplanes will be analyzed. Firstly, the breaking can be caused by *rotation* of the hyperplanes with respect to one another. Secondly, the hyperplanes can be *translated* with respect to another. Thirdly, a composite mechanism of the previous two mechanisms may occur; this will however not be analyzed in this chapter.

The rotation-based breaking can be monitored by the angle φ between hyperplanes in a cluster. An analysis of rotation-based breaking of clustered hyperplanes and a discussion of the relation with the encounter of temporary minima follows in section 8.5. Section 8.6 presents illustrative examples of rotation-based breaking of a cluster of hyperplanes.

Translation-based breaking assumes that the hyperplanes move away from each other in a parallel way, and can be monitored by the distance between the hyperplanes. The analysis and brief discussion of the translation-based breaking mechanism is presented in section 8.7 and an example is presented in section 8.8.

8.4 Rotation-based breaking

In this section, the mechanism behind rotation-based breaking of a cluster of hyperplanes will be analyzed. Because β_1^h corresponds to the norm of the weight vector component perpendicular to the attractor hyperplane and $|\underline{\Delta W}^{\varepsilon}|$ is related to the norm of the weight vector component in parallel to the attractor hyperplane, with (8.15a) and (8.15b), the angle between the two hyperplanes of the two neurons in the first layer is

$$\varphi = 2 \cdot atan\left(\frac{|\underline{\Delta W}^{\varepsilon}|}{2\,\beta_1^h}\right). \tag{8.20}$$

With the backpropagation learning algorithm, the weight vector increments during training are small. Therefor, the increment of the angle between the two hyperplanes for one adaptation is:

$$\Delta\varphi = \frac{\partial\varphi}{\partial|\underline{\Delta W}^{\varepsilon}|}\cdot\Delta|\underline{\Delta W}^{\varepsilon}| + \frac{\partial\varphi}{\partial\beta_1^h}\cdot\Delta\beta_1^h \tag{8.21}$$

The partial derivatives on the right hand side of (8.21) can directly be derived from (8.20):

$$\frac{\partial\varphi}{\partial|\underline{\Delta W}^{\varepsilon}|} = \frac{4\,\beta_1^h}{4\,\beta_1^{h^2} + |\underline{\Delta W}^{\varepsilon}|^2}, \tag{8.22}$$

$$\frac{\partial\varphi}{\partial\beta_1^h} = \frac{-4\,|\underline{\Delta W}^{\varepsilon}|}{4\,\beta_1^{h^2} + |\underline{\Delta W}^{\varepsilon}|^2}. \tag{8.23}$$

The two increments $\Delta|\underline{\Delta W}^{\varepsilon}|$ and $\Delta\beta_1^h$ have to be derived from the backpropagation learning algorithm. With (8.19), for the increments of $\underline{W}_{1,j}^{\varepsilon}$ during learning of one training example,

$$\Delta\underline{W}_{1,j}^{\varepsilon,p} = \eta(D-Y)f'(\alpha_{2,1}^h\cdot\beta_{2,1}^h+\alpha_{2,1}^{bias}\,\beta_{2,1}^{bias})\frac{\beta_{2,1}^h}{\sqrt{2}}\cdot f'(\alpha_1^h\cdot\beta_1^h+\alpha_1^{bias}\,\beta_1^{bias}+\underline{W}_{1,j}^{\varepsilon}\cdot\underline{U}_1^F)\,\underline{U}_1^F \tag{8.24}$$

where $f'(\cdot)$ is the first derivative of the output of the transfer function with respect to its input. The increment of $|\underline{\Delta W}^{\varepsilon}|$ for one training example, $\Delta|\underline{\Delta W}^{\varepsilon}|^p$, can be obtained from (8.13):

$$\Delta|\underline{\Delta W}^{\varepsilon}|^p = \Delta|\underline{W}_{1,1}^{\varepsilon,p} - \underline{W}_{1,2}^{\varepsilon,p}|. \tag{8.25}$$

A first order approximation of (8.25) is

$$\Delta|\underline{\Delta W}^{\varepsilon}|^p = \eta(D-Y)f'(\alpha_{2,1}^h\cdot\beta_{2,1}^h+\alpha_{2,1}^{bias}\,\beta_{2,1}^{bias})\frac{\beta_{2,1}^h}{\sqrt{2}}\cdot f''(\alpha_1^h\cdot\beta_1^h+\alpha_1^{bias}\,\beta_1^{bias})\left(\underline{U}_1^F\cdot\underline{\Delta W}^{\varepsilon}\right)|\underline{U}_1^F|, \tag{8.26}$$

where $f''(\cdot)$ denotes the second derivative of the output of the transfer functions of the neurons with respect to the input. This first order approximation is sufficiently accurate as long as the angle between the hyperplanes is small.

The increment of the difference vector $\underline{\Delta W}^{\varepsilon}$ can be calculated by summation of (8.26) over all training examples. For reasons of mathematical notation and operation, we use the integral equivalent of the sum

$$\Delta|\,\underline{\Delta W}^{\varepsilon}| = \int_{p} \Delta|\,\underline{\Delta W}^{\varepsilon}|^{p}\; P(p)\; \mathrm{d}p\,. \tag{8.27}$$

In (8.27), $P(p)$ denotes the probability density function of the pth training example. Substitution of (8.26) in (8.27) results in

$$\Delta|\,\underline{\Delta W}^{\varepsilon}| = \int_{\alpha_1^h} C_{\underline{W\varepsilon}}(\alpha_1^h) \int_{\underline{U}_1^F} |\underline{U}_1^F|\,\left(\underline{U}_1^F \cdot \underline{\Delta W}^{\varepsilon}\right) P(\alpha_1^h, \underline{U}_1^F)\; d\underline{U}_1^F \;\; d\alpha_1^h \tag{8.28}$$

$$\text{with: } C_{\underline{W\varepsilon}}(\alpha_1^h) = \eta(D-Y) f'(\alpha_{2,1}^h \beta_{2,1}^h + \alpha_{2,1}^{bias}\beta_{2,1}^{bias}) \frac{\beta_{2,1}^h}{\sqrt{2}} f''(\alpha_1^h\beta_1^h + \alpha_1^{bias}\beta_1^{bias})\,.$$

Note that the desired response, or target response, D in (8.28) is different for the two classes to be classified. Similarly, the adaptation of β_1^h of the two neurons in the first layer can be derived. The increment $\Delta\beta_{1,j}^{h,p}$ on one training example is

$$\begin{aligned}\Delta\beta_{1,j}^{h,p} = {} & \eta(D-Y) f'(\alpha_{2,1}^h \cdot \beta_{2,1}^h + \alpha_{2,1}^{bias}\beta_{2,1}^{bias}) \frac{\beta_{2,1}^h}{\sqrt{2}} \\ & f'(\alpha_1^h \cdot \beta_1^h + \alpha_1^{bias}\beta_1^{bias} + \underline{W}_{1,j}^{\varepsilon} \cdot \underline{U}_1^F)\,\alpha_1^h\;.\end{aligned} \tag{8.29}$$

The $\Delta\beta_1^h$ over the total training set can be approximated by

$$\Delta\beta_1^h \approx \int_{\alpha_1^h}\int_{\underline{U}_1^F} C_{\beta 1}(\alpha_1^h)\; P(\alpha_1^h, \underline{U}_1^F)\; d\underline{U}_1^F \;\; d\alpha_1^h \tag{8.30}$$

$$\text{with: } C_{\beta 1}(\alpha_1^h) = \eta(D-Y) f'(\alpha_{2,1}^h \beta_{2,1}^h + \alpha_{2,1}^{bias}\beta_{2,1}^{bias}) \frac{\beta_{2,1}^h}{\sqrt{2}} \cdot f'(\alpha_1^h\beta_1^h + \alpha_1^{bias}\beta_1^{bias})\,\alpha_1^h$$

By substitution of (8.22) and (8.23) into (8.21), it can be derived that

$$\Delta\varphi \approx \sin(\varphi) \cdot \left[\frac{\Delta|\,\underline{\Delta W}^{\varepsilon}|}{|\,\underline{\Delta W}^{\varepsilon}|} - \frac{\Delta\beta_1^h}{\beta_1^h}\right], \tag{8.31}$$

where the adaptation of β_1^h and $\underline{\Delta W}^{\varepsilon}$ are given by (8.30) and (8.28). In the next sections, the interpretation of the results denoted by (8.31) are discussed. Simulations that illustrate the effect of (8.31) on the learning behavior are presented in section 8.5.

8.4.1 Discussion

The effect of (8.31) on the learning behavior of the two-layer neural network will now be explained. It is important to note that the normalized increments of $|\underline{\Delta W}^{\varepsilon}|$ and of β_1^h at the right hand side of (8.31) are independent of φ. Firstly, the two normalized increments on the right hand side of (8.31) and secondly the dynamics of the angle φ during training will be discussed.

Normalized increment of $|\underline{\Delta W}^{\varepsilon}|$

The first term between brackets in (8.31) is the normalized increment of the $\underline{\Delta W}^{\varepsilon}$ vector over all training examples,

$$\Delta|\underline{\Delta W}^{\varepsilon}|_{norm} = \frac{\Delta|\underline{\Delta W}^{\varepsilon}|}{|\underline{\Delta W}^{\varepsilon}|}. \tag{8.32}$$

It can be derived from (8.26) that adaptation in response to one training example can result in either an increment or a decrement of $|\underline{\Delta W}^{\varepsilon}|$. For the training examples for which the integrand in (8.27) equals zero, the superscript 0 will be used. For the examples $\underline{U}_1^0$:

$$\Delta|\underline{\Delta W}^{\varepsilon}|^{p}(\underline{U}_1^0) = 0 \tag{8.33}$$

There are two trivial and one non-trivial solutions for training examples satisfying (8.33). One trivial solution is found for examples for which the derivative of the transfer function of the neuron in the second layer is zero. Another trivial solution of (8.33) is found for examples which are already classified correctly; for these training examples the weight adaptation is zero.

Non-trivial examples satisfying (8.33) are those for which the first derivatives of the transfer function of the neurons in the first layer are identical, see (8.24) and (8.25). This is true as a good approximation for all training examples for which the second derivatives of the transfer function of the neurons in the first layer equal zero, see (8.26). When using odd transfer functions in the neurons in the first layer of the network, it follows from (8.11) and (8.26) that these non-trivial examples are located in the attractor hyperplane of the neurons in the first layer,

$$f(-x)=2f(0)-f(x) \quad \rightarrow \quad \underline{U}_1^0 \cdot \underline{W}_1^{ATT} = 0\,. \tag{8.34}$$

Furthermore, it can be derived from (8.26) that

$$\frac{|D-f(\sqrt{2}\beta_{2,1}^h f(\alpha_1^h\beta_1^h+\alpha_1^{bias}\beta_1^{bias})+\alpha_{2,1}^{bias}\beta_{2,1}^{bias})|}{|D-f(\sqrt{2}\beta_{2,1}^h f(0)+\alpha_{2,1}^{bias}\beta_{2,1}^{bias})|} > 1 \Leftrightarrow \Delta|\underline{\Delta W}^{\varepsilon}|^{p} > 0. \tag{8.35}$$

Relation (8.35) denotes that training examples which lie on the "wrong" side of the attractor hyperplane from a classification point of view, have a positive contribution to $\Delta|\underline{\Delta W}^{\varepsilon}|$. Training examples which lie on the "correct" side of the attractor hyperplane tend to reduce the norm of $\underline{\Delta W}^{\varepsilon}$.

Training examples that result in weight adaptation are bounded by both the training set boundaries and by the training examples for which $(D-Y)=0$. With (8.28), (8.32) and (8.35), it can be derived that

$$\Delta|\underline{\Delta W}^{\varepsilon}|_{norm}(t+\tau) > \Delta|\underline{\Delta W}^{\varepsilon}|_{norm}(t) \qquad \tau>0 \tag{8.36}$$

under condition of small initial weights (in terms of the definition in section 8.4).

Normalized increment of β_1^h

The second term between brackets in (8.31) is the normalized increment of β_1^h over all training examples,

$$\Delta\beta_1^h{}_{norm} = \frac{\Delta\beta_1^h}{\beta_1^h}. \tag{8.37}$$

During training, (8.37) will be positive as long as (8.18) is satisfied, assuming positive weights associated with neurons in the first layer for the neuron in the second layer. Using small initial weights, $\Delta\beta_1^h$ decreases during training for several reasons. Firstly, the number of correctly classified training examples increases during training, which results in an average decrease of the mean adaptation. Secondly, the mean adaptation on not yet correctly classified training examples decreases. For the normalized increment,

$$\Delta\beta_1^h{}_{norm}(t) > \Delta\beta_1^h{}_{norm}(t+\tau) \geq 0 \qquad \tau>0 \tag{8.38}$$

The angle φ during training

Under assumption of small initial weights, three significantly different cases are distinguished during training. These cases are determined by the sign of the sum of the normalized increments of β_1^h and $|\underline{\Delta W}^{\varepsilon}|$.

- The first situation occurs if the expression between brackets in (8.31) is positive at the beginning of the training,

$$\left[\frac{\Delta|\underline{\Delta W}^{\varepsilon}|}{|\underline{\Delta W}^{\varepsilon}|} - \frac{\Delta\beta_1^h}{\beta_1^h}\right] > 0 \qquad t\geq 0. \tag{8.39}$$

  In this case, the sum will remain positive because of (8.36) and (8.38). With (8.36) and (8.38), it follows from (8.35) that the situation reflected by (8.39) only occurs if the classification made by the network for small φ is very poor.

  From (8.31), it follows that there exists a set of φ for which φ is constant. These angles will be denoted as *invariant angles*. The invariant angles are given by

$$\varphi = n\pi \qquad n \in \{0,1,2,\ldots\}. \tag{8.40}$$

  With (8.31), (8.39) and (8.40), it follows that angles with even n are unstable and that angles with odd n are stable. According to (8.31), for any initial angle that is not an even multiple of

π, φ goes asymptotically to the nearest odd multiple of π. Note that if the two-layer neural net is in an equilibrium, the two hyperplanes of the neurons in the first layer coincide.

$$\begin{cases} \varphi = 2n\,\pi & n \in \{0, 1, 2, \ldots\} & \Rightarrow \text{unstable} \\ \varphi = (2n{+}1)\,\pi & n \in \{0, 1, 2, \ldots\} & \Rightarrow \text{stable} \end{cases} \tag{8.41}$$

- In the second possible situation for the training of the two-layer neural network, the square brackets term in (8.31) is negative throughout the training,

$$\left[\frac{\Delta|\underline{\Delta W}^{\varepsilon}|}{|\underline{\Delta W}^{\varepsilon}|} - \frac{\Delta\beta_1^h}{\beta_1^h}\right] < 0 \qquad t \geq 0\,. \tag{8.42}$$

With (8.36) and (8.38), it follows from (8.35) that the classification made by the neural network is good. From (8.31) it follows that in this case the invariant angles between the two hyperplanes are also given by (8.41). However, the stable and unstable angles are interchanged with respect to (8.41),

$$\begin{cases} \varphi = 2n\,\pi & n \in \{0, 1, 2, \ldots\} & \Rightarrow \text{stable} \\ \varphi = (2n{+}1)\,\pi & n \in \{0, 1, 2, \ldots\} & \Rightarrow \text{unstable} \end{cases} \tag{8.43}$$

In this situation, the angle between the two hyperplanes decreases asymptotically to an even multiple of π during the entire training; the weight vector attractors of the neurons in the first layer coincide. Because the two hyperplanes of the neurons in the first layer tend to coincide during training, the classification of the training set will be carried out using redundant neurons in the first layer of the neural network; i.e. the two neurons make almost identical classifications.

- The last possible learning behavior is the most interesting with respect to the derivation of several properties concerning the learning. In this last case, the term between brackets in (.31) is negative in the beginning of the training, and becomes positive during training.

$$\begin{cases} \left[\frac{\Delta|\underline{\Delta W}^{\varepsilon}|}{|\underline{\Delta W}^{\varepsilon}|} - \frac{\Delta\beta_1^h}{\beta_1^h}\right] \leq 0 & t \leq \tau_{crit} \\ \left[\frac{\Delta|\underline{\Delta W}^{\varepsilon}|}{|\underline{\Delta W}^{\varepsilon}|} - \frac{\Delta\beta_1^h}{\beta_1^h}\right] > 0 & t > \tau_{crit} \end{cases} \tag{8.44}$$

It is clear that in this case, the learning behavior sequentially surpasses the two previously mentioned cases. In the beginning of the training, the angle between the hyperplanes converges towards an even multiple of π. After a certain point in the training, however, the angle between the hyperplanes is driven to an odd multiple of π. This means that the weight vector attractors of the neurons in the first layer coincide for $t \leq \tau_{crit}$ and are non-coinciding elsewhere.

Discussion of negative elements in $\underline{B}_{2,1}^{h}$

In the analysis, all elements of $\underline{B}_{2,1}^{h}$ were assumed to be positive. In general, however, the elements of $\underline{B}_{2,1}^{h}$ can be either positive or negative. The effect of negative weights associated with neurons in the first layer on the learning behavior of two-layer neural networks is small; a negative element in $\underline{B}_{2,1}^{h}$ means that for the associated neuron in the first layer the weight vector points in the opposite direction compared to a positive weight:

$$\underline{W}_{2,1}[j] \cdot \beta_{1,j}^{h} > 0 \tag{8.45}$$

Hence, for neurons in the first layer connected to a negative weight, both $\beta_{1,j}^{h}$ and $\underline{W}_{1,j}^{\varepsilon}$ have opposite signs compared to neurons whose outputs are connected to positive weights. The effect on the preceding analysis is that the stable and unstable equilibria of the hyperplanes in (8.41) and (8.43) are interchanged for every non-positive element in $\underline{B}_{2,1}^{h}$. Note that also the expression for $\underline{\Delta W}^{\varepsilon}$ in (8.14) must be changed and that the signs of $\beta_{1,j}^{h}$ must be compensated for.

Another effect, not analyzed in this chapter, is that the training time needed to obtain a certain level of performance for mixed positive and negative elements in $\underline{B}_{2,1}^{h}$ may differ from the training time for the case of only positive elements.

Temporary minima

In this section, it is shown that the learning behavior depicted in (8.44) leads towards the encounter of a temporary minimum. To happen upon the learning behavior of (8.44), the training set must satisfy a number of conditions. Firstly, it follows from (8.35), (8.36), (8.38) and (8.44) that the training set needs to be non-linearly separable. Secondly, the performance of the neural network with clustered hyperplanes must be non-optimum in order to 'break' the cluster of hyperplanes. Thirdly, the performance of the neural network with clustered hyperplanes in the first layer must be good enough to ensure that the situation depicted in (8.44) is encountered; a very poor classification of the training set using clustered hyperplanes results in (8.39).

While training a network for $\tau \leq \tau_{crit}$ in (8.44), the angle between the hyperplanes corresponding to the neurons in the first layer decreases monotonically. This means that the network builds up a redundancy of the neurons in the first layer. The classification of the training set made by the network approaches in this phase the classification made by a network with redundancy. The analysis as presented so far requires that (8.18) be satisfied which is true for either small weights or a small angle between the weight vectors; this latter includes a small angle φ between the hyperplanes. It can easily be shown that (8.18) is satisfied for $\tau \leq \tau_{crit}$ in (8.44) assuming small initial weights.

In the subsequent phase, $t > \tau_{crit}$ in (8.44), the angle φ increases monotonically which indicates gradual abolishing of the redundancy in the first layer. The analysis as presented in this chapter remains valid as long as the angle φ remains small, see (8.18). After (8.18) is no longer satisfied, the hyperplanes corresponding to the neurons in the first layer make a significantly different classification; the input space has been subdivided, or partitioned [12], into two parts each classified by one neuron in the first layer. In this case, the analysis of the learning behavior of a two-layer neural net can be continued by making new vector decompositions for the two neurons in the first layer; the directions of the $\underline{B}_{1j}^{h}$ component vectors are then related to the part to be classified by the corresponding neuron and is hence different for both neurons in the first layer. Due to abolition of the redundancy of the neurons in the first layer, the classification of the training examples can be improved significantly. This vanishing of redundancy results in 'escaping' from the temporary minima [16]. It can directly be seen from the term $\sin(\varphi)$ in (8.31) that abolishing the redundancy, i.e. increasing of the angle φ, is generally a slow process. This explains why neural nets may 'stick' in a temporary minimum during training for a relatively long time.

8.5 Rotation-based based breaking: an illustrative example

In this section, the equations describing the dynamics of the two-layer neural network during training, (8.28), (8.30) and (8.31), will be used to describe the learning behavior for a simple classification problem that results in subsequently clustering and rotation-based breaking of the formed cluster of hyperplanes.

Training set to be learnt
To demonstrate the conclusions of sections 8.3 and 8.4, we consider the learning behavior of a two layer feed-forward neural net for classification of two classes in two-dimensional space. The classes in this training set A are indicated in figure 8.7 by the two dotted areas.

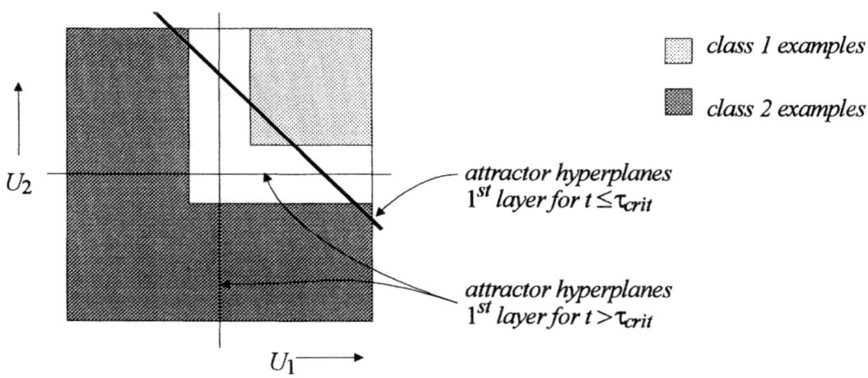

Fig. 8.7 Training set A and the corresponding attractor hyperplanes of the first-layer neurons; thick line for $t \leq \tau_{crit}$ and thin lines for $t > \tau_{crit}$

Note that the all neurons in the neural network have a three-dimensional input and weight vector. The input vectors for the neurons in the first layer have a two external input elements and one bias input element, see (8.1). The neuron in the second layer has two inputs from the neurons in the first layer and one bias input, see (8.2).

Initialization

Before starting the training of the neural network, the weights of the neurons in the network have to be initialized. It is assumed that the initial weights are small enough to allow for linearization of the neural network at the beginning of training. According to section 8.4, under this condition, the weight vectors of the neurons in the first layer adapt almost identically in the beginning of training. Furthermore, the elements in $\underline{W}_{2,1}$ which are not related to the bias input are also adapting almost identically in this phase.

To illustrate this, figure 8.8 presents the simulated dynamics of the weights (excluding the bias related weights) in weight space. These simulation results are obtained using a neural network simulation program that trained the specified neural net on the training set of figure 8.7

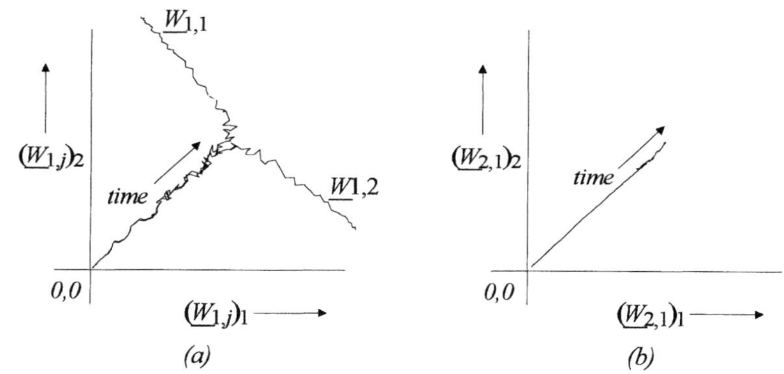

Fig. 8.8 Simulated dynamics of the (non-bias) weights of the neurons in weight space

Figure 8.8a clearly illustrates that in the beginning of training using very small initial weights, the non-bias related vector component in $\underline{W}_{1,j}$, i.e. the vectors $\underline{W}^{r}_{1,j}$, adapt identically. After a certain training time, the so-formed cluster of hyperplanes abruptly 'breaks' into two hyperplanes which make a significantly different classification. The direction in which $\underline{W}^{r}_{1,j}$ adapts in weight space is dependent on the training set for both the phase of training in which hyperplanes are clustered and for the phase of training in which the neurons in the first layer make a significantly different classification.

Figure 8.8b shows that the non-bias related weights in $\underline{W}_{2,1}$, i.e. the elements of $\underline{W}^{r}_{2,1}$, adapt identically during the training. In the beginning of training, this is due to the fact that very small initial weight are used which allows for linearization. After 'breaking' of the cluster of hyperplanes, the non-bias related weights in $\underline{W}_{2,1}$ continue to adapt identically which is due to symmetry reasons, i.e. the fact that for the used training set the two parts that are classified by the non-clustered hyperplanes are equally important for the classification of the total training set.

Phase portrait

Under the condition of small initial weights, the learning behavior of the two-layer neural network is deterministic according to (8.28), (8.30) and (8.31). Note that the order of presentation of the training examples is not important; the network adapts according to (8.6) in which the weight adaptation is gradient descent for the squared error over all training examples.

Because of the deterministic behavior of the learning having global learning and negligibly small initial weights, the β^{h}_{kj}, β^{bias}_{kj} and $|\underline{W}^{\varepsilon}_{kj}|$ of all neurons in the neural network are strongly correlated. The state of the neural network, which is a function of the β^{h}_{kj}, β^{bias}_{kj} and $|\underline{W}^{\varepsilon}_{kj}|$ of all neurons can therefore be expressed in for example β^{h}_{1} and φ. By doing so, a two-dimensional phase portrait [4] of the learning behavior of the two-layer neural network on the training set of figure 8.7 can be constructed; this phase portrait is given in figure 8.9. Because equations (8.28), (8.30) and (8.31) have been derived under condition of (8.18), this phase portrait is valid only after the linearization phase (see section 8.3) and until the hyperplanes make a significantly different classification. For subsequent phases of training, a new vector decomposition is to be made for every cluster of redundant neurons and hence new phase portraits result.

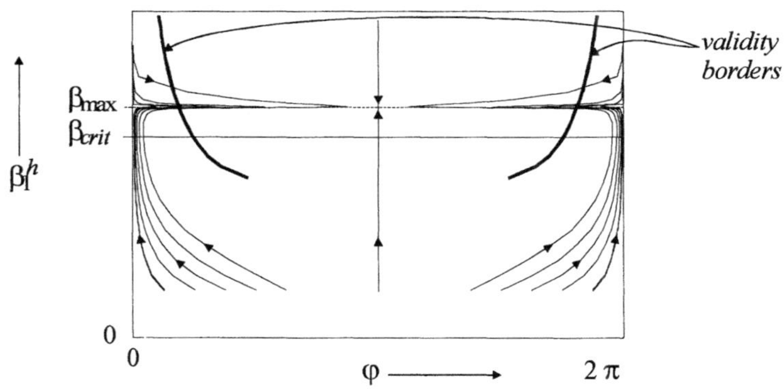

Fig. 8.9 Phase portrait for the learning behavior of a two-layer neural network: rotation-based breaking of clustered hyperplanes

In figure 8.9, the phase trajectories start for values of β^{h}_{1} that marginally do not allow linearization of the network which corresponds to the begin situation for clustering and breaking behavior according to (8.31). In the figure the initial angles φ have been chosen in a large range for reasons of illustration.

The β^{h}_{1} moves asymptotically towards β_{max} during training, which corresponds to the value of β^{h}_{1} obtained for a redundant neural network after an infinite training. For non-redundant networks, the angle φ moves towards an even integer multiple of π for $\beta^{h}_{1} < \beta_{crit}$, with β_{crit} corresponding to τ_{crit} in (8.44). After β_{crit} has been reached, φ moves towards an odd multiple of π.

The phase portrait is valid as long as (8.18) is satisfied. The validity borders in figure 8.9 mark the borders below which condition (8.18) is satisfied. As soon as a phase trajectory crosses one of these validity borders, a new vector decomposition has to be made.

Phase portrait assuming major simplifications

The phase portrait in figure 8.9 has been constructed by using the equations in sections 8.3 and 8.4, which is very time consuming. However, with some major simplifications, it is possible to very simply construct an approximating phase portrait: assume that β_1^h approaches β_{max} exponentially and assume that the normalized increment of the norm of the $\underline{\Delta W}^{\varepsilon}$ vector is a linearly increasing function of β_1^h. In this way the phase portrait can be constructed using

$$\begin{cases} \Delta\varphi = A\sin(\varphi)\left[\beta_1^h - \beta_{crit} - \dfrac{\beta_{max} - \beta_1^h}{\beta_1^h} + \dfrac{\beta_{max} - \beta_{crit}}{\beta_{crit}}\right] \\ \Delta\beta_1^h = \beta_{max} - \beta_1^h \end{cases}$$

(with A a scalar scale value) which generally gives an acceptably accurate approximation of the actual phase portrait for training sets that result in rotation-based breaking of clusters of hyperplanes. With this set of difference equations, among others the changing of the distribution of the examples from which the network learns and the changing adaptation on these training examples are simplified significantly. However, one can easily (and fast) get a good impression of the neural net's learning behavior using this set of difference equations.

Learning behavior for $0 < \beta_{crit} < \beta_{max}$
It follows from (8.35) that in the case of the non-linearly separable training set A, β_{crit} is smaller than β_{max}. For sets with $\beta_{crit} < \beta_{max}$, the angle φ goes asymptotically to an even multiple of π for $t \leq \tau_{crit}$ in (8.43). During continued training, $t > \tau_{crit}$, φ tends to converge to an odd multiple of π. This means that for the first layer with redundant neurons at $t = \tau_{crit}$, the redundancy is lifted for $t > \tau_{crit}$.

Learning behavior for $\beta_{crit} > \beta_{max} > 0$
For some training sets, particularly those with linearly separable classes, β_{crit} will be higher than β_{max}, see (8.35). In this case, on condition of small initial weights, the network will tend to build up redundancy in the first layer during the entire training. In figure 8.10, an example of a (linearly separable) training sets for which $\beta_{crit} > \beta_{max}$ is given.

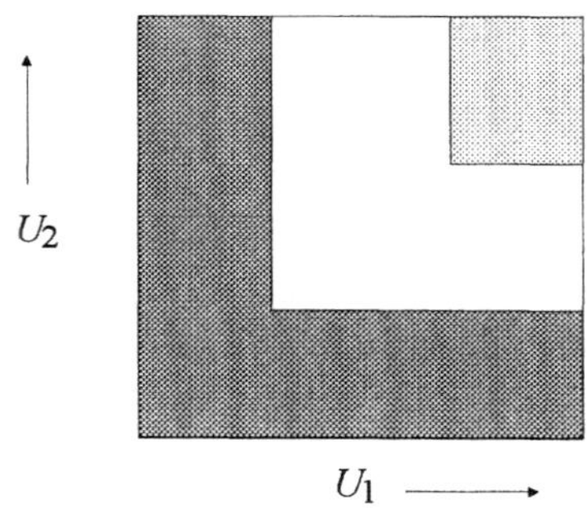

Fig. 8.10 Training set B with $\beta_{crit} > \beta_{max}$

For linearly separable classes, the inequality $\beta_{crit} > \beta_{max}$ is not a problem for the classification to be learnt as any neural network with redundant neurons in the first layer can classify the separable training set correctly.

The starting conditions for the analysis, which are satisfied when using small enough initial weights, will generally not be satisfied when using a priori knowledge. In case of proper use of a priori knowledge in initial weights, the hyperplanes corresponding to the neurons in the first layer may have different attractor hyperplanes throughout the training. In that case no redundancy will be built up in the beginning of training.

Phase portrait allowing negative elements in $\underline{B}_{2,1}^{h}$

As discussed in section 8.3.1, the stable and unstable invariant angles in (8.41) and (8.43) may interchange for negative elements in $\underline{B}_{2,1}^{h}$. Directly after initializing the neural network using negligibly small weights for all neurons in the neural network, $\underline{B}_{2,1}^{h}$ can not be defined.

The direction in which $\underline{B}_{2,1}^{h}$ points after a short training period is now determined by two competing weight vector attractors. One attractor attracts to a state for which the stable angles are at even multiples of π, see (8.41) corresponding to an even number of negative weights in $\underline{W}_{2,1}$. The other attractor attracts to equilibrium states for which odd multiples of π are stable; this corresponds to an odd number of negative weights in $\underline{W}_{2,1}$.

Therefore, as soon as the weight vectors in the network are attracted to a stable angle, the resulting phase portrait is either the phase portrait shown in figure 8.9 or a phase portrait identical to the one in figure 8.9 but shifted in the φ direction over π due to interchanging of the stable and unstable angles.

Required training time

The training time required to obtain a certain level of performance for a particular training set has been calculated with the help of (8.30) and (8.31) using a special purpose computer program. For training set A, the values of $\beta_{1,n}^{h}$ for the neurons in the first layer have been calculated as a function of the number of training examples. In figure 8.11, these (coinciding) curves have been plotted (thick lines).

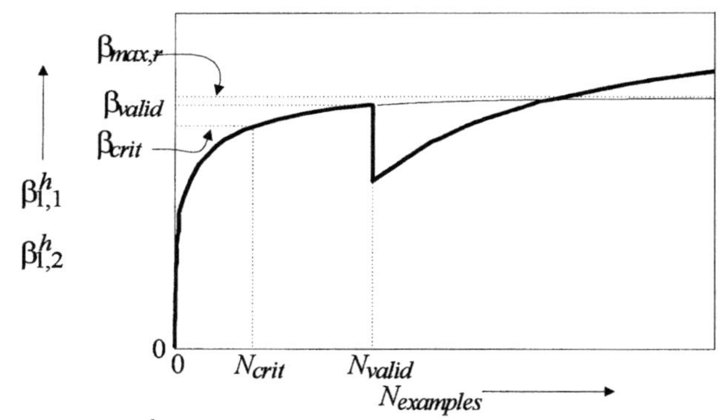

Fig. 8.11 $\beta_{1,n}^{h}$ as a function of the number of training examples

In figure 8.11, β_{crit} corresponds to the value of β_{1}^{h} at τ_{crit} in (8.44) and β_{valid} corresponds to the value of β_{1}^{h} for which (8.18) is marginally satisfied, see also figure 8.9. As soon as β_{1}^{h} equals β_{valid}, the two neurons in the first layer start to make a significantly different classification. At this point in training, a new vector decomposition of the weight vectors of the neurons in the first layer has to be made to be able to continue the analysis. This new vector decomposition causes the drop of the $\beta_{1,n}^{h}$ curves which is by a factor $\cos(\theta)$ where θ is the rotation of the $\underline{B}_{1,n}^{h}$ vectors at β_{valid}, for this particular example $\theta=45^{o}$. After making a significantly different classification of the two hyperplanes, the values of $\beta_{1,n}^{h}$ continue to increase to generally larger values than $\beta_{max,r}$. The coincidence of $\beta_{1,1}^{h}$ and $\beta_{1,2}^{h}$ after the breaking of the cluster is due to the symmetry of the particular training set.

The mean squared error

An indication for the performance of a feed-forward neural network is the mean squared error of the output signal of the network calculated over all training examples [8],[16],[19],[20] .
For the specified network and training set A, the mean squared error has both been calculated and simulated as a function of the number of training examples. In figure 8.12, the calculated and simulated MSE versus training time curves and the angle between the hyperplanes φ are plotted; note the logarithmic y-axis for φ.

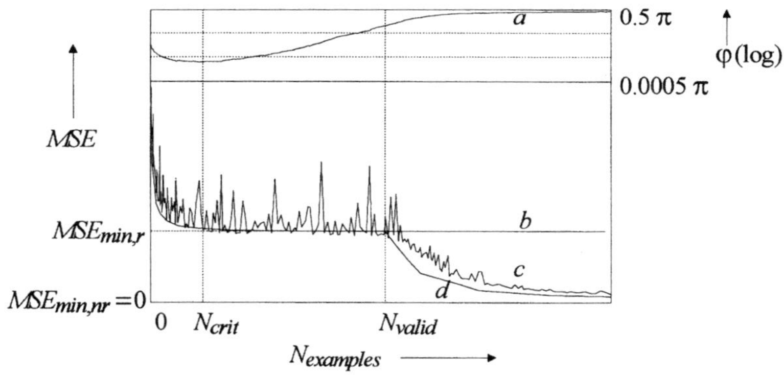

Fig. 8.12 (a) The angle φ,
(b) the calculated MSE for a redundant network, and
(c) the simulated and (d) calculated MSE for a non-redundant network
all as a function of the number of training examples

For a network with redundant neurons in the first layer, it is not possible to classify all training examples correctly. Therefore, the mean squared error of network with redundant neurons does not reach zero but will stick in a temporary minimum. Assuming proper initialization, the network will however lift the redundancy during training.
Assuming proper initialization, the MSE versus training time curve in the beginning of the training is approximately identical to the curve obtained for a network with redundant neurons in the first layer because the network builds up redundancy. The MSE will therefore approach the MSE corresponding to a network with redundant neurons in the first layer. During training, the redundancy will slowly be abolished as soon as $\beta_{1}^{h} \approx \beta_{crit}$, marked with N_{crit} in figure 8.12. This slow annihilation of the redundancy can be noticed in the MSE versus training time curve as a temporary minimum, see figure 8.12.

The analysis as presented in this section is valid until the angle φ is so large that (8.18) is no longer satisfied. In figure 8.12, this point is marked with N_{valid}. After N_{valid} has been reached, the hyperplanes corresponding to the two neurons in the first layer make a significantly different classification; the input space has been subdivided, or partitioned [12] over the two neurons in the first layer. The analysis of the learning behavior can now be continued after making a new vector decomposition with the $\underline{B}_{1,j}^{h}$ vectors related to the (two) attractor hyperplanes corresponding to the two parts of the subdivided input space.

Due to the removal of the redundancy of neurons in the first layer of the network, the classification of the training examples can be improved considerably. The MSE therefore decreases relatively quickly after N_{valid}. Note that because training set A can be classified with a two-layer neural net having two neurons in the first layer, the MSE will approach zero after N_{valid}.

In the temporary minimum, it might be concluded from the MSE versus training time curve that the increase of the performance of the total network approaches zero. However, as can be derived from curve (a) in figure 8.12 and as shown by the analyses in this chapter, a fundamental performance improving process takes place in the interval between N_{crit} and N_{valid}. This is observed in the adaptation of the angle φ, while it is not noticeable in the MSE versus training time curve.

From (8.35), (8.38) and (8.44), it can be derived that the better the classification by a network with redundant neurons, the lower MSE_{min} and the higher N_{crit} will be. Therefore, for relatively good classification of a non-linearly separable training set by a network with redundant neurons, the network 'sticks' in a temporary minimum for a long (training) time.

For training sets for which the classification made using a reduced network is poor, the network "sticks" in a temporary minimum only during a relatively short training time. In the MSE versus training time curve, this 'sticking' phenomenon may not even be recognized.

8.6 Translation-based breaking

For some training sets, the declustering of the hyperplanes is not rotation-based as analyzed in section 8.4 but in a parallel manner, called translation based breaking. In this section, mechanism behind translation-based breaking of clustered hyperplanes is analyzed. The analysis again assumes very small initial weights (as defined in section 8.3) and follows the same line as the analysis of the rotation-based breaking of clustered hyperplanes in section 8.4.

For the translation-based breaking of clustered hyperplanes, it will be assumed that the clustered hyperplanes move away from each other in a parallel manner. In this way, the norms of the $\underline{W}^{\varepsilon}$ vector components are assumed to be zero. The distance between the hyperplanes is now determined by the $\beta^{h}_{1,j}$ and by the $\beta^{bias}_{1,j}$ of the two neurons with clustered hyperplanes. It is straight forward to derive that (under condition of a zero norm for the $\underline{W}^{\varepsilon}$ vector component) the distance between a hyperplane and the origin of input space is given by

$$d\left(\underline{0}, hyperplane_{1,j}\right) = \left|\frac{\beta^{bias}_{1,j}}{\beta^{h}_{1,j}}\right|.$$

With (8.14), the distance between the hyperplanes of the two neurons in the first layer (in parallel with respect to each other), is then

$$dist = \left|\frac{\Lambda\beta^{bias}_{1}}{\beta^{h}_{1}}\right|. \tag{8.46}$$

During training, the weight vectors of all neurons in the neural network adapt. As a direct result, the distance between the two hyperplanes in parallel also adapt during training. Analogous to the derivation in section V, the adaptation of the distance *dist* is:

$$\Delta dist \approx dist \cdot \left[\frac{\Delta\Lambda\beta^{bias}_{1}}{\Lambda\beta^{bias}_{1}} - \frac{\Delta\beta^{h}_{1}}{\beta^{h}_{1}}\right]. \tag{8.47}$$

Equations (8.29) and (8.30) describe the adaptation of β^{h}_{1}. The adaptation of $\Lambda\beta^{bias}_{1}$ on the pth training example is then analogous to (8.26) to (8.31) given by:

$$\Delta\Lambda\beta^{bias,p}_{1} = \eta\,(D-Y)\,f'\left(\alpha^{h}_{2,1}\,\beta^{h}_{2,1} + \alpha^{bias}_{2,1}\,\beta^{bias}_{2,1}\right)\frac{\beta^{h}_{2,1}}{\sqrt{2}} \cdot \tag{8.48}$$

$$f''\left(\alpha^{h}_{1}\,\beta^{h}_{1} + \alpha^{bias}_{1}\,\beta^{bias}_{1}\right)\left(\alpha^{bias}_{1}\,\Lambda\beta^{bias}_{1}\right)\alpha^{bias}_{1} \;.$$

The average adaptation $\Lambda\beta^{bias}_{1}$ is then

$$\Delta\Lambda\beta^{bias}_{1} = \int_{p} \Delta\Lambda\beta^{bias,p}_{1}\; P(p)\, dp\,. \tag{8.49}$$

Discussion

The effect of (8.47) on the learning behavior is comparable to the effect of (8.31). As a direct result, three possible situations can occur during training similar to the discussions in section 8.4. This section first discusses some properties of the first term between brackets in (8.47) and then discusses the three possible types of learning behavior that may occur with (8.47). The discussions follow the same line as section 7.4.

Normalized increment of $\Lambda\beta_1^{bias}$

The first term between brackets in (8.47) is the normalized increment of $\Lambda\beta_1^{bias}$ over all training examples. It follows from (8.48) that the adaptation of $\Lambda\beta_1^{bias}$ in response to one training example is either positive or zero or negative. For the training examples for which the integrand in (8.49) equals zero, the superscript 0 will be used. For the examples $\underline{U}_1^0$:

$$\Delta|\Lambda\beta_1^{bias}|^p(\underline{U}_1^0) = 0 \tag{8.50}$$

There are a few trivial and one non-trivial solutions for training examples satisfying (8.50). Analogous to the discussion of (8.33), trivial solutions are firstly the examples for which no adaptation takes place, secondly the situation in which $\alpha_1^{bias}=0$ and thirdly the situation in which the clustered hyperplanes are coinciding (i.e. $\Lambda\beta_1^{bias}=0$).

Non-trivial examples satisfying (8.50) are training examples for which the first derivatives of the transfer function of the neurons in the first layer are identical. This is as a good approximation the case for all training examples for which the second derivatives of the transfer function of the neurons in the first layer equal zero, see (8.48). For odd transfer functions in the neurons in the first layer of the network, non-trivial solutions of (8.50) are located in the attractor hyperplane of the neurons in the first layer. It can be derived that for these training examples

$$\frac{|D-f(\sqrt{2}\beta_{2,1}^h f(\alpha_1^h\beta_1^h+\alpha_1^{bias}\beta_1^{bias})+\alpha_{2,1}^{bias}\beta_{2,1}^{bias})|}{|D-f(\sqrt{2}\beta_{2,1}^h f(0)+\alpha_{2,1}^{bias}\beta_{2,1}^{bias})|}>1 \Leftrightarrow \Delta|\Lambda\beta_1^{bias}|^p>0. \tag{8.51}$$

In words, training examples which lie on the "wrong" side of the attractor hyperplane from a classification point of view, have a positive contribution to $\Delta|\Lambda\beta_1^{bias}|$ and tend therefore to "break" the cluster to translation of the clustered hyperplanes. Training examples which lie on the "correct" side of the attractor hyperplane tend to increase the clustering.

Three possible type of learning behavior

The effect of (8.47) on the learning behavior is comparable to the effect of (8.31). Similarly to the discussion in section 8.4, three possible situations can occur during training.

- Firstly, the term between brackets in (8.47) is positive throughout the training. In this case, the hyperplanes diverge (in parallel with respect to each other) at any point in training. In this case, the classification of the training set with clustered hyperplanes of neurons in the first layer is very poor. The stable and unstable invariant distances are then given by:

$$\begin{cases} dist = 0 & \Rightarrow \text{unstable} \\ dist \to \infty & \Rightarrow \text{stable} \end{cases} \tag{8.52}$$

Note that as soon as the distance *dist* between the two hyperplanes is so large that (8.18) is no longer satisfiede, a new vector decomposition is to be made for both neurons in the first layer.

- Secondly, the term between brackets in equation (8.47) is negative during training. It may be evident that the hyperplanes of the two neurons in the first layer contract towards the same attractor throughout training. In this case, the classification of the training set with one cluster of hyperplanes (virtually one neuron in the first layer) is good. Similar to (8.41) and (8.43), the stable and unstable invariant distances are interchanged with respect to (8.52).
- Thirdly, the term between brackets in (8.47) may be negative in only the beginning of training and becomes positive at a certain point in training. This situation is comparable to the situation depicted by (8.44) in section 8.4:

$$\begin{cases} \left[\dfrac{\Delta \Lambda\beta_1^{bias}}{\Lambda\beta_1^{bias}} - \dfrac{\Delta\beta_1^h}{\beta_1^h}\right] \leq 0 & t \leq \tau_{crit} \\ \left[\dfrac{\Delta \Lambda\beta_1^{bias}}{\Lambda\beta_1^{bias}} - \dfrac{\Delta\beta_1^h}{\beta_1^h}\right] > 0 & t > \tau_{crit} \end{cases} \tag{8.53}$$

In this case, a temporary minimum is encountered during training in a way similar to the way described for rotation-based breaking of clustered hyperplanes in section 8.4.

Discussion of negative weights in $\underline{B}_{2,1}^h$

It was assumed for the analysis in this section that the elements in $\underline{B}_{2,1}^h$ are positive. For negative elements in this unit vector, the hyperplanes of associated neurons in the first layer are rotated 180 degrees with respect to a positive element. Due to the calculation of the absolute value in (8.46), the sign of any element in $\underline{B}_{2,1}^h$ does not have any effect for the translation-based breaking mechanism. Note however that when allowing negative signs in $\underline{B}_{2,1}^h$ results in a different expression for $\Lambda\beta_1^{bias}$ and that also compensations for the signs of $\beta_{1,j}^h$ must be included.
Another difference resulting from negative elements in $\underline{B}_{2,1}^h$ (not analyzed in this chapter) is that the training time required to reach a certain level of performance may be changed.

Temporary minima

It has been shown in section 8.4 that the learning behavior depicted by (8.31) and (8.44) leads towards the encounter of a temporary minimum, which was due to subsequently building up and removing redundancy in the first layer of a two-layer feed-forward neural network.

The learning behavior depicted in (8.47) and (8.53) is strongly related to the behavior of (8.31) and (8.44); only minor differences exist. To avoid major redundancy in the text of this chapter, we do not repeat an extensive discussion of the occurrence of temporary minima for translation-based breaking of clustered hyperplanes.

8.7 Translation-based breaking: an illustrative example

In this section, simulation results that illustrate the translation-based breaking of clustered hyperplanes are presented. The training set used in this example is again two-dimensional and suited for classification using a two-layer neural network with two neurons in the first layer: the XOR-problem (shown in figure 8.13) has been selected.

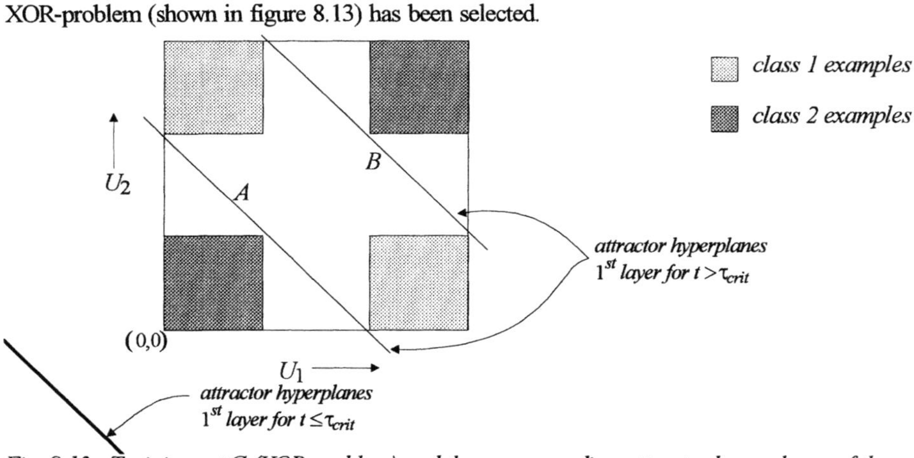

Fig. 8.13 Training set C (XOR problem) and the corresponding attractor hyperplanes of the neurons in the first layer; thick line for $t \leq \tau_{crit}$ and thin lines for $t > \tau_{crit}$

In the very beginning of training, the hyperplanes of the two neurons in the first layer form a cluster. This cluster is positioned in input space in such a way that the solution of the neural network for the training set is optimum. With the origin of input space in the lower left corner of the training set (indicated by the coordinates (0,0)), the cluster is positioned outside the training set, in the lower left corner of figure 8.13. During training, the hyperplanes continue to move towards each other for $t \leq \tau_{crit}$.

The clustered hyperplanes move away from each other in a parallel way for $t > \tau_{crit}$, and move towards their final position in input space. This final position is indicated by the two thin lines in figure 8.13.

Phase portrait

Similar as for the learning behavior with rotation-based breaking of clustered hyperplanes, a phase portrait can be constructed of the learning behavior with translation-based breaking. In the phase portrait, the distance between the hyperplanes in parallel is plotted as a function of $\beta_{1,n}^{h}$. The simulated phase portrait of the learning behavior of a single-output two-layer neural network with two first-layer neurons on the training set of figure 8.13 is shown below.

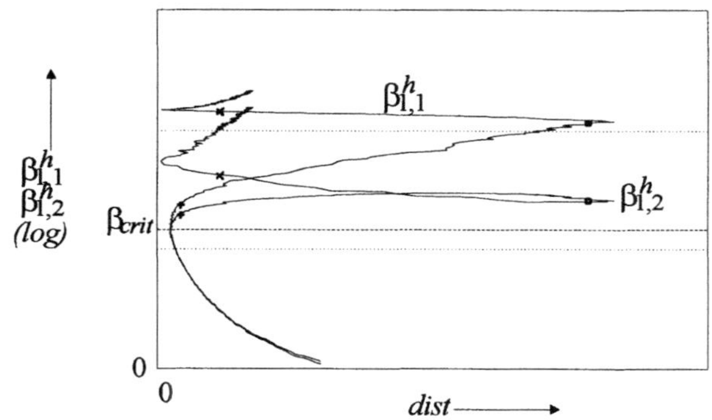

Fig. 8.14 Phase portrait for the learning behavior of a two-layer neural network: translation-based breaking of clustered hyperplanes

In figure 8.14, the distance *dist* has been plotted as a function of both $\beta_{1,1}^{h}$ and $\beta_{1,2}^{h}$ because after breaking the cluster at $t=\tau_{crit}$ the adaptation of $\beta_{1,1}^{h}$ and $\beta_{1,2}^{h}$ is different. Note that after (8.18) is no longer satisfied, the weight vector attractors of the two neurons in the first layer are significantly different which results in divergence of the two trajectories in the phase portrait. To compare the trajectories after breaking of the cluster, a number of markers have been included in figure 8.14: identical markers at the two trajectories correspond to specific points in training. At a certain point in training and after declustering of the hyperplanes, the distance between the two hyperplanes decreases rapidly and then increases to its final value. This is explained as follows: directly after declustering of the hyperplanes one of the hyperplanes moves towards attractor hyperplane A in figure 8.13. The other hyperplanes moves after a certain time towards attractor hyperplane B in figure 8.13 for which it has to pass the hyperplanes that attracts to attractor A. This passing process causes the rapid decrease and increase of the *dist* variable in figure 8.14 after declustering.

Phase portrait assuming major simplifications
Calculation of a phase portrait for translation-based type of learning, using the equations in sections 8.3, 8.4 and 8.6 is very time consuming. With some major simplifications, it is possible to very simply construct an approximating phase portrait. Proposed simplifications are: assume that β_1^h approaches β_{max} exponentially and assume that the normalized increment of $\Delta\beta_1^{bias}$ is a linearly increasing function of β_1^h. In this way the phase portrait can be constructed using

$$\begin{cases}\Delta dist = A\, dist\left[\beta_1^h - \beta_{crit} - \dfrac{\beta_{max} - \beta_1^h}{\beta_1^h} + \dfrac{\beta_{max} - \beta_{crit}}{\beta_{crit}}\right] \\ \Delta\beta_1^h = \beta_{max} - \beta_1^h \end{cases}$$

(with A a scalar scale value) which generally gives an acceptably accurate approximation of the actual phase portrait for training sets that result in translation-based breaking of clusters of hyperplanes. With this set of difference equations, one can easily get a satisfactorily good impression of the neural net's learning behavior.

The mean squared error
It follows from figure 8.14 that the neural network 'sticks' in a temporary minimum for relatively small β_1^h. This means that the temporary minimum is encountered in the beginning of training, just after the phase in which the network can be linearized (see section 8.3). In figure 8.15, the simulated mean squared error (MSE) of the network is plotted as a function of the number of presented training examples.

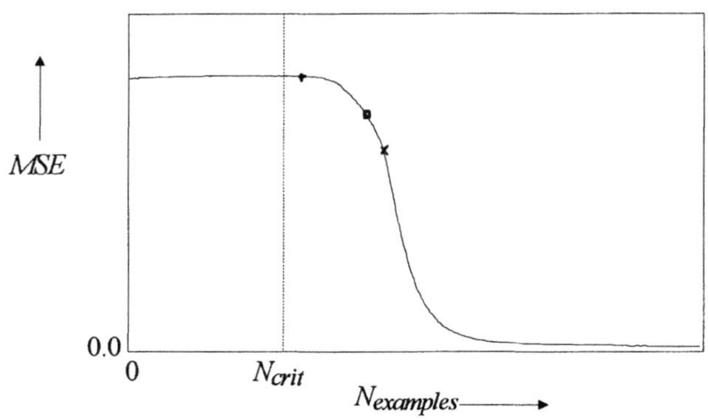

Fig. 8.15 Simulated MSE for the XOR-problem; markers corresponding to the markers in the phase portrait in figure 8.14.

The markers that were plotted in figure 8.14 are also included in the MSE versus training examples curve. Note that for this particular training set, the major part of the 'sticking' time is required to build up the redundancy in the first layer of the two-layer neural network, whereas for the rotation-based mechanism the redundancy-lift required most 'sticking' time.

8.8 Extension towards larger networks

In the analyses and examples in the previous sections of this chapter, the number of neurons in the first layer was limited to two. It will be shown by induction, that the analyses of sections 8.3, 8.4 and 8.6 are also valid for neural nets with more than two neurons in the first layer.

For elucidation, assume a first layer with an arbitrary number of neurons. Assuming sufficiently small initial weights, the network can be linearized in the very beginning of training. During this phase, the weight vectors of all neurons will adapt almost identically, if again assuming only positive elements in $\underline{B}^h_{2,1}$. The starting conditions for the analyses in sections 8.3, 8.4 and 8.6 will hence be satisfied in the very beginning of training for neural nets with an arbitrary number of neurons in the first layer.

As all neurons in the first layer adapt according to (8.24), (8.30) and (8.48), it follows that all hyperplanes converge towards the same position in input space for $\tau \leq \tau_{crit}$ in (8.44) or (8.53). At the end of this phase of training, the first layer will be highly redundant.

In the subsequent phase, for $\tau > \tau_{crit}$ in (8.44) or (8.53), the cluster of redundant neurons effectively 'breaks' into two clusters of redundant neurons, each cluster consisting of at least one neuron. This 'breaking' agrees with the mechanisms derived for the two-neuron case in sections 8.4 and 8.6. After (8.18) is no longer satisfied, the two clusters classify a significantly different but neighboring part of the training set.

The learning behavior of each cluster after making a significantly different classification, can be analyzed by making a new vector decomposition using the (new) attractor hyperplanes corresponding to the analyzed cluster. For each cluster, the analyses as presented in section 8.4 and 8.6 hold when neglecting the effects of other clusters on the analyzed cluster. Note that only neglecting the effect of *neighboring* clusters introduces a small error because only a small part of the examples classified by neighboring clusters is also classified by the analyzed cluster. It can be concluded that for networks with an arbitrary number of neurons in the first layer, the network builds up redundancy in the beginning of training, and iteratively abolishes the redundancy partially during further training by iteratively 'breaking' of clusters of redundant neurons into smaller clusters.

An example of the learning behavior of a neural network with more than two neurons in the first layer is presented in the following figure. Different phases in learning are represented by line segments between the black dots; training starts at the leftmost dot and proceeds towards the right.

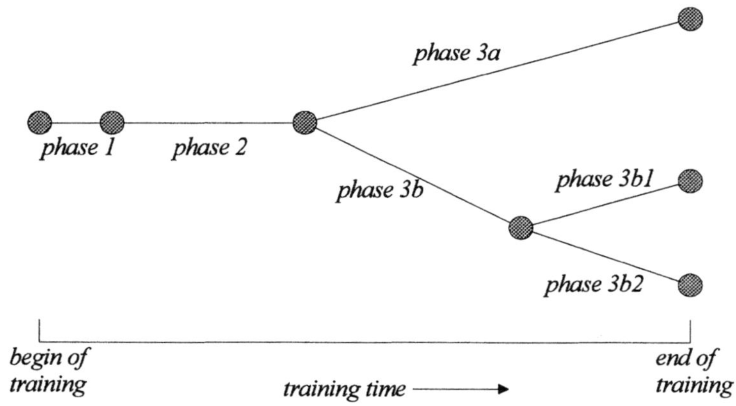

Fig. 8.16 Example of the phases of the learning behavior of a neural network with more than two neurons in the first layer

In phase 1, the weights are small enough to allow linearization and consequently redundancy in the first layer is built up; this phase has been discussed in section 8.3.

In phase 2, linearization of the network is not allowed because neurons are generally activated outside the middle (approximately linear) region. In this phase of training, redundancy in the first layer is first increased and then gradually abolished by tending to subdivide the cluster of neurons into two smaller clusters. This subdivision can be either rotation-based (as presented in section 8.4) or translation-based (presented in section 8.6). The slow process of tending to 'break' into multiple clusters may be recognized in the MSE versus training time curve as a temporary minimum.

In phases 3a and 3b, the mutual redundancy is partially vanished, there are two clusters of neurons that classify a significantly different part of the input space. For both clusters, a new attractor hyperplane must be defined that corresponds to the 'subdivided' training set. Strong redundancy exists in each cluster and the analysis can now be done for each cluster separately using the new attractor hyperplanes. As illustrated by the line segments 3b1 and 3b2, clusters may iteratively 'break' into smaller clusters.

Discussion of "very small" initial weights

For the analysis presented in this chapter, *very small* initial weights have been assumed. In neural network training, this assumption is generally satisfied when no a priori knowledge has been included in the weights. It has been shown in the previous subsection that for an appropriate training set (not separable by clustered hyperplanes in the first layer while better separable after 'breaking' the cluster) the condition of very small initial weights leads to the encounter and 'escape' of temporary minima.

However, if the initial weights are not *very small* according to the definition in section 8.3, another mechanism leading towards temporary minima may occur. Although a thorough analysis of this mechanism is beyond the scope of this chapter, this situation is discussed briefly.

It has been shown that under condition of very small initial weights, the redundancy is build up by clustering hyperplanes. If the condition of very small weights is not fulfilled, temporary minima may result from two types of redundancy. Firstly, one of the two clustering mechanisms derived for the case of very small initial weights may be present; phase 3b in figure 8.16 corresponds to a situation in which not small "initial" weights ("initial" here means at the beginning of the analysis of phase 3b) leads towards a temporary minimum due to clustering and 'breaking' of the cluster.

Secondly, the neural network may only significantly adapt the weight vector of a part of the neurons in the first layer. This means that other neurons in the first layer are either not actively used in the classification or are trained such that they have a constant output which will act as an extra bias input for the neuron in the second layer. In the first case, the norm of the weight vector of these neurons initially remains relatively small and the associated weight in $\underline{W}_{2,1}$ also remains small. To 'escape' an encountered minimum, a not-actively used neuron may become activated. In the latter case, the neurons have a constant response, resulting from positioning the hyperplane relatively far away from the training set while the norm of the weight vector corresponding to the hyperplane is large; these neurons will not actively be used in the classification. This last type of minimum is described by Wessels and Barnard [24].

8.9 Conclusions

Temporary minima are usually recognized from the mean squared error (MSE) versus training time curve as a part in which the MSE is approximately constant for a long time after initial training. Better understanding of the mechanism behind this type of minimum is important because this may lead to finding methods for minimizing training time of two-layer neural nets. It has been shown in this chapter that the temporary minima result from subsequently building up redundancy in the first layer and removal of this redundancy. Removing the redundancy requires a relatively long training time in which the MSE sticks at a level corresponding to a network with redundant neurons. As soon as the redundancy is (partially) abolished, the MSE curve drops relatively steep to a significant lower level corresponding to a less redundant neural network.

8.10 References

[1] S.Akaho, and S.Amari, "On the Capacity of Three-Layer Networks", in Proc. IJCNN, San Diego, vol. 3, pp. 1-6, 1990

[2] M.Arai, "Mapping Abilities of Three-Layer Neural Networks", in Proc. IJCNN, Washington D.C., vol. 1, pp. 419-423, 1989

[3] M.T.Gately, and P.A.Penz, "A geometrical Overview of Neural Networks", TI Technical Journal, pp. 4-15, 1990

[4] J.Guckenheimer, and P.Holmes, "Nonlinear Oscillations, Dynamical Systems, and Bifurcations of Vector Fields", New York: Springer-Verlag, 1983

[5] H.Guo, and S.B.Gelfand, "Analysis of Gradient Descent Learning Algorithms for Multilayer Feedforward Neural Networks", IEEE Trans. Circuits and Systems, vol. 38, pp. 883-894, 1991

[6] S.J.Hanson, and D.J.Burr, "Minkowski-r Back-Propagation: Learning Connectionist Models with Non-Euclidian Error Signals", in *Neural Information Processing Systems*, ed. D.Z.Anderson, AIP, New York, pp. 348-357, 1988

[7] T.M.Heskes, E.T.P.Slijpen, and B.Kappen, "Learning in neural networks with local minima", Physical Review A, vol. 46, pp. 5221-5231, 1992

[8] Y.Hirose, K.Yamashita, and S.Hijiya "Back-Propagation Algorithm Which Varies the Number of Hidden Units", Neural Networks, vol. 4, pp. 61-66, 1991

[9] K.Hornik, M.Stinchcombe, and H.White, "Universal Approximation of an Unknown Mapping and Its Derivatives Using Multilayer Feedforward Networks", Neural Networks, vol.3, pp. 551-560, 1990

[10] S.C.Huang,and Y.F.Huang,"Bounds on the Number of Hidden Neurons in Multilayer Perceptrons", IEEE Trans. Neural Networks, vol. 2, pp. 47-55, 1991

[11] B.Irie, and S.Miyake, "Capabilities of Three-layered Perceptrons", in Proc. IJCNN, San Diego, vol 1, pp. 641-648, 1988

[12] P.Liang, "Design Artificial Neural Networks Based on the Principle of Divide-and-Conquer", in Proc. IEEE ISCAS, Singapore, pp. 1319-1322, 1991

[13] R.P.Lippmann , "An Introduction into Computing with Neural Nets", IEEE ASSP Magazine, vol.4, pp. 4-22, 1987

[14] K.G.Mehrotra, C.K.Mohan, and S.Ranka, "Bounds on the Number of Samples Needed for Neural Learning", IEEE Trans. Neural Networks, vol. 2, pp. 548-558, 1991

[15] M.L.Minsky, and S.A.Papert, "Perceptrons, An Introduction to Computational Geometry", Expanded Edition, Cambridge, MA: MIT Press, 1988

[16] A.F.Murray, "Analog VLSI and Multi-layer Perceptrons- Accuracy, Noise, and On-Chip Learning", in Proc. 2nd Int. Conf. on Microelectronics for Neural Networks, Munich, pp. 27-34, 1991

[17] A.F.Murray, "Analogue noise-enhanced Learning in Neural Network Circuits", Electronic Letters, vol. 27, pp. 1546-1548, 1991

[18] D.B.Parker, "Learning Logic", Technical Report 47, MIT, 1985

[19] D.E.Rumelhart, G.E.Hinton, and R.J.Williams, "Learning internal representations by error propagation", in *Parallel Distributed Processing: Exploration in the microstructure of cognition* (Vol.1). Cambridge, MA: MIT Press, 1986

[20] B.J.Sheu, "VLSI Neurocomputing with Analog Programmable Chips and Digital Systolic Array Chips", in Proc. IEEE ISCAS, Singapore, pp. 1267-1270, 1991

[21] E.D.Sontag, and H.J.Sussman "Backpropagation Separates Where Perceptrons Do, Neural Networks, vol. 4, pp. 243-249, 1991

[22] H.J.Sussmann, "Uniqueness of the Weights for Minimal Feedforward Nets with a Given Input-Output Map", Neural Networks, vol. 5, pp. 589-593, 1992

[23] P.J.Werbos, "Backpropagation: Past and Future", in Proc. IJCNN, San Diego, vol. 1, pp. 343-353, 1988

[24] L.F.A.Wessels and E.Barnard, "Avoiding False Local Minima by Proper Initialization of Connections", IEEE tr. Neural Networks, vol. 3, no. 6, pp. 899-905, 1992

[25] D.Woods, "Back and Counter Propagation Abberations", in Proc. IJCNN, San Diego, vol. 1, pp. 473-479, 1988

Chapter 9

Biases and Unipolar Input signals for Two-Layer Neural Networks

9.1 Introduction

In this chapter, a few topics for two-layer feed-forward neural network training are discussed. Sections 9.2 and 9.3 analyses are concerned with the effect of the magnitude of the (constant) bias signal on the learning behavior of two-layer nets. Normally the bias inputs of the neurons in a neural network are set to unity. Sections 9.2 and 9.3 show that the value of this bias input for respectively the first and second layer has a direct effect on the learning behavior of the network. Using the centre-of-gravity of (parts of) the training set (see section 4.2.1), an optimal bias signal can be determined for every neuron in the network. This optimal bias is generally different for all neurons.

It is known from literature that first-layer neurons may be trained as to produce constant responses, independently of the input vector [9]. As a result of this, the second-layer neuron is virtually extra biased. Section 9.4 analyses this effect and presents a solution.

In chapter 4, it has been argumented that using two-quadrant multipliers in hardware realizations is efficient from an implementation point of view. To be able to use two-quadrant multipliers, the input signals of any neuron need however to be unipolar. Section 9.5 discusses the consequences on training time when translating the input signals in input space in such a way that unipolar input signals are obtained.

9.2 Effect of the first layer's bias-input signal on learning

It was shown in chapter 3 that in single-layer feed-forward neural networks, the magnitude of the bias input signal of the neurons has a direct effect on the (average) weight adaptation. Furthermore, it was shown in sections 4.2.2 and 4.2.3 that for *single neurons,* the adaptation of the weight vector is optimal if:

- $\alpha^{bias} \gg |\underline{U}^{CG} - \alpha^{bias}\underline{B}^{bias}|$
  if the learning rate η is small enough to ensure stability during training, or

- $|\alpha^{bias,opt}| = \sqrt{-\underline{U}^{F,const^2} + |\alpha^{ATT}|\sqrt{(N_{in}-1)\hat{U}^2 - |\underline{U}^{F,const}|^2}}$
  if the learning rate is scaled as to obtain the maximum adaptation in the optimum direction

Instead of scaling the actual bias-input-signal in order to optimize the learning behavior, one may choose to modify the learning rate associated with only the bias-input signal. It has been shown in section 3.4 that scaling one element in the input vector by a factor x is identical to scaling the learning rate η of only that element by x^2. Therefore, another way to optimize the learning behavior of single-layer nets is to properly increase the learning rate of only the bias-related weight.
For single-output *two-layer neural networks*, it can be expected that the magnitude of the bias input signal also affects the learning behavior. In this section, a short analysis of the effect of the magnitude of the bias input signal, applied to neurons in the first layer of a two-layer network, on training time is presented.

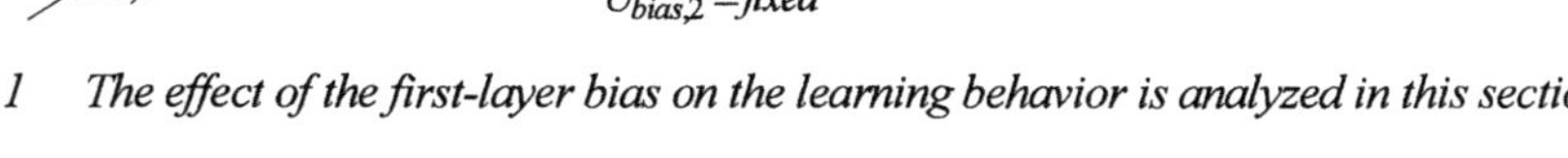

Fig. 9.1 The effect of the first-layer bias on the learning behavior is analyzed in this section

In the analysis, results obtained in previous chapters will be used; these results are briefly discussed in section 9.2.1. The analysis itself is given in sections 9.2.2 and 9.2.3. In section 9.2.4, simulation results illustrate the derived relations and the results of this section are summarized in 9.2.5.

9.2.1 Learning behavior: a recapitulation

It has been shown in chapter 8 that in two-layer neural network training, three fundamentally different learning phases can be distinguished:

- In the first phase of training, directly after starting the training using very small initial weights, the network can be linearized which is explained in section 8.3.1. In this phase, first-layer hyperplanes start to contract towards each other. Because this first phase in training occurs only in the very beginning of training, the analyses in this chapter are applied to only the second and third phase.
- In the second phase of training, the clustered first-layer hyperplanes continue to contract towards each other which corresponds to learning for $t<\tau_{crit}$ in (8.44). During training in this second phase, the networks converges towards a minimum; this minimum can be global, local or temporary. If a global or local minimum is encountered, the increase of performance decreases to zero and the third phase is not reached. In case of convergence towards a temporary minimum, the neural network eventually enters the third phase which in turn results in escaping from the temporary minimum.
- In the third phase of training, a cluster of hyperplanes breaks into generally two smaller clusters of hyperplanes in order to improve the performance, which situation corresponds to $t>\tau_{crit}$ in (8.44). This "breaking" of clustered first-layer hyperplanes into more clusters corresponds to leaving a temporary minimum. After this escape, the resulting (smaller) clusters of hyperplanes enter a new phase 2, see also section 8.8.

It was derived in chapter 8 that under assumption of symmetry, this contracting and breaking of clustered hyperplanes in the first layer of a two-layer feed-forward neural network can be either rotation-based or translation-based. For the rotation-based breaking, the angle φ between two clustered hyperplanes adapts as

$$\Delta\varphi \approx \sin(\varphi)\cdot\left[\frac{\Delta|\Lambda W^{\varepsilon}|}{|\Lambda W^{\varepsilon}|}-\frac{\Delta\beta_1^h}{\beta_1^h}\right], \tag{9.1}$$

where the distance between two clustered hyperplanes with the translation-based mechanism is given by a similar equation:

$$\Delta dist \approx dist\cdot\left[\frac{\Delta\Lambda\beta_1^{bias}}{\Lambda\beta_1^{bias}}-\frac{\Delta\beta_1^h}{\beta_1^h}\right]. \tag{9.2}$$

The sign of the factor between brackets in (9.1) and (9.2) determines whether the first-layer hyperplanes enforce the clustering or "break" the clustering; the clustered hyperplanes contract for a negative factor and split up for a positive factor.

It has been shown in chapter 6 that the slow process of breaking up of clustered hyperplanes is associated with 'sticking' in a so-called temporary minimum in the MSE versus training time curve. This can be explained by noting that the part of the training set that initially is classified by one cluster of hyperplanes will after breaking up of the cluster be classified by more clusters. This results in effectively subdividing the original part of the training set in two smaller parts for which the classification can be improved.

Reintroduction of the centre-of-gravity

In section 4.2.1, the centre-of-gravity of a training set was introduced for single-layer neural networks. This centre-of-gravity is defined as the point in input space for which the adaptation of the weight vector of the neuron classifying the training set would be *maximum* in the local optimal $\underline{B}^h$ direction, while the deviation from this local optimal direction would be *minimum* if the origin of input space coincides with the centre-of-gravity.

For two-layer feed-forward neural networks, it may be clear that every neuron in the net has its specific centre-of-gravity, which lies in the part of the input space classified by that neuron. The centre-of gravity corresponding to the nth neuron in the first layer of a two-layer network will be denoted as $\underline{U}_{1,n}^{CG}$ with

$$\underline{U}_{1,n}^{CG} = \alpha_{1,n}^{ATT}\, \underline{B}_{1,n}^{h} + \alpha^{bias}\, \underline{B}_{bias} + \underline{U}_{1,n}^{F,const} \tag{9.3}$$

Note that for all neurons *in a cluster*, the centre-of-gravity is identical. After breaking of a cluster, the centres-of-gravity of the newly formed clusters of hyperplanes are significantly different.

9.2.2 First layer's bias input versus adaptation in the $\underline{B}^h$ direction

The adaptation of the weight vectors in either layer of a two-layer neural network is slow. As a consequence, the weight vector of the neuron in the second layer can by approximation be assumed fixed or quasi-static during a limited time interval. With this assumption, the adaptation $\beta_{1,n}^h$ on the pth training example is

$$\Delta\beta_{1,n}^{h,p} = \eta\,(D^p - Y^p)\,f'(\underline{W}_2\cdot\underline{U}_2^p)\,\frac{\beta_{2,1}^h}{\sqrt{2}}\,f'(\alpha_1^{h,p}\,\beta_1^h + \alpha_1^{bias}\beta_1^{bias})\,\alpha_{1,n}^{h,p}. \tag{9.4}$$

Note that generally every neuron in the first layer of a two-layer neural network has its specific local optimal direction $\underline{B}_{1,n}^h$. To derive an expression for the average adaptation of $\beta_{1,n}^h$, a decomposition of $\alpha_{1,n}^h$ and $\underline{U}^F$ into constant and a zero-mean part is required (this is similar to the decompositions in sections 3.5 and 3.6) is required:

$$\alpha_{1,n}^h = \alpha_{1,n}^{h,v} + \alpha_{1,n}^{ATT} \tag{9.5a}$$

$$\underline{U}_{1,n}^F = \underline{U}_{1,n}^{F,v} + \underline{U}_{1,n}^{F,const}. \tag{9.5b}$$

The average adaptation of the vector component of $\underline{W}_{1,n}$ pointing in the $\underline{B}_{1,n}^h$ direction can be derived in a similar manner as in section 3.6. Under assumption of a quasi-static weight vector of the second-layer's neuron, the resulting equation is

$$\Delta\beta_{1,n}^h = \frac{\eta'}{P}\cdot\frac{\beta_{2,1}^h}{\sqrt{2}}\sum_{p=1}^{P}(D^p - Y^p)\,f'(\underline{W}_2\cdot\underline{U}_2^p)\,f'(\underline{W}_{1,n}\cdot\underline{U}_1^p)\,\alpha_{1,n}^{h,v,p} - \frac{\alpha_{1,n}^{ATT}\,\underline{U}_{1,n}^{F,const}\cdot\left(\overline{\Delta\underline{W}_{1,n}^{\varepsilon}}\right)_{W\varepsilon}}{{\alpha_1^{bias}}^2 + |\underline{U}_{1,n}^{F,const}|^2 + {\alpha_{1,n}^{ATT}}^2} \tag{9.6}$$

$$\text{with } \eta' = \eta\left(\frac{{\alpha_1^{bias}}^2 + |\underline{U}_{1,n}^{F,const}|^2}{{\alpha_1^{bias}}^2 + {\alpha_{1,n}^{ATT}}^2 + |\underline{U}_{1,n}^{F,const}|^2}\right)$$

An equation describing the adaptation of the moving average part of the $\underline{W}_{1,n}^{\varepsilon}$ vector will not be given in this chapter; it is however comparable to the equation derived for single-layer neural networks in section 3.6.

Equation (9.6) gives a direct relation between the adaptation of $\Delta\beta^h_{1,n}$, the bias input signal applied to the neuron, and the distance between the attractor hyperplane and the origin of input space. The equation assumes a quasi-constant weight vector of the neuron in the second layer of the two-layer neural network. As this assumption is satisfied at any point in training for small enough learning rates η, (9.6) can be used throughout the training. Denoting the time at which $\beta^h_{1,n}$ has reached a specific value as $\tau_{\beta_{1,n}}$ it follows from (9.6) that

$$\tau_{\beta_{1,n}} = \Phi(\alpha_1^{bias}) \quad with \quad \frac{\partial\Phi(\alpha_1^{bias})}{\partial\alpha_1^{bias}} < 0. \tag{9.7}$$

9.2.3 Relation between first layer's bias input and temporary minima

For the learning behavior of two-layer neural networks, not only the adaptation of the weight vector in the $\underline{B}^h_{1,n}$ direction is important; a large part of training time of two-layer neural nets is usually spent in so-called temporary minima, see for example figure 8.1. Before reaching a temporary minimum, hyperplanes corresponding to neurons in the first layer form clusters. Each cluster of hyperplanes contains hyperplanes which approximately coincide.

Escaping from temporary minima is accomplished by 'breaking' clusters in multiple clusters of hyperplanes. Each newly formed cluster now classifies a part of the training examples which were classified by the original cluster. As 'breaking' of clusters generally requires a long (training) time, temporary minima can be recognized in MSE versus training time curves as regions in which the MSE is virtually constant for a certain limited training time.

Clustering and declustering

The formation and breaking of clusters can proceed according to two related mechanisms: the rotation-based and the translation-based mechanism. The clustering-declustering mechanism is given by (9.1) and (9.2). It has been derived in chapter 8 that if the training set can be classified better after 'breaking' the clustered hyperplanes into two (or more) clusters, there is a β_{crit} that satisfies

$$\begin{cases} \beta^h_{1,n} < \beta_{crit} & \Rightarrow \quad clustering \\ \beta^h_{1,n} \geq \beta_{crit} & \Rightarrow \quad declustering \end{cases}. \tag{9.8}$$

The second term between brackets in (9.1) and (9.2) tends to enforce the clustering of hyperplanes because this term is (except when using very large initial weights, see figure 8.8.9) always is positive. When including the minus-sign this means that this term always decreases either φ in (9.1) or always decreases *dist* in (9.2).

With (9.6), it follows that this $\Delta\beta^h_1/\beta^h_1$ term in (9.1) and (9.2) is a monotonously increasing function of the bias input applied to the neurons in the first layer. In formula:

$$\Delta\beta^h_1/\beta^h_1 = \Gamma(\alpha_1^{bias}) \quad with \quad \frac{\partial\Gamma(\alpha_1^{bias})}{\partial\alpha_1^{bias}} > 0. \tag{9.9}$$

For the clustering and declustering behavior of first-layer hyperplanes, also the first term between brackets in (9.1) and (9.2) is important. These terms are $\Delta|\underline{\Delta W}^{\varepsilon}|/|\underline{\Delta W}^{\varepsilon}|$ for the rotation-based breaking depicted in (9.1) and $\Delta\Lambda\beta_1^{bias}/\Lambda\beta_1^{bias}$ for the translation-based breaking mechanism in (9.2). It follows from (8.35) and (8.51) that training examples that lie on the "wrong" side of the clustered hyperplanes from a classification point of view result in an increase of these terms. Other training examples result either in no weight adaptation or in a decrease:

$$\begin{cases} \underline{U}_1^p \text{ on } \textbf{wrong} \text{ side of clustered hyperplanes} & \Rightarrow \quad \Delta|\underline{\Delta W}^{\varepsilon}| > 0 \\ \underline{U}_1^p \text{ on } \textbf{correct} \text{ side of clustered hyperplanes} & \Rightarrow \quad \Delta|\underline{\Delta W}^{\varepsilon}| \leq 0 \end{cases} \tag{9.10}$$

and

$$\begin{cases} \underline{U}_1^p \text{ on } \textbf{wrong} \text{ side of clustered hyperplanes} & \Rightarrow \quad \Delta\Lambda\beta_1^{bias} > 0 \\ \underline{U}_1^p \text{ on } \textbf{correct} \text{ side of clustered hyperplanes} & \Rightarrow \quad \Delta\Lambda\beta_1^{bias} \leq 0 \end{cases} \tag{9.11}$$

An analysis of the dependence of the $\Delta|\underline{\Delta W}^{\varepsilon}|/|\underline{\Delta W}^{\varepsilon}|$ and $\Delta\Lambda\beta_1^{bias}/\Lambda\beta_1^{bias}$ terms on the bias signal applied to the first-layer neurons is presented below.

In two-layer neural networks, the clustered hyperplanes may coincide with the decision boundary of the total neural network; in formula with $\underline{U}_1^b$ in the decision boundary of the neural net,

$$\underline{W}_{1,n} \cdot \underline{U}_1^b = 0 \qquad \Leftrightarrow \qquad Y_{net}(\underline{U}_1^b) = f(0). \tag{9.12}$$

For this (special) case, the number of examples that are positioned at the 'wrong' side of the clustered hyperplanes is minimum. However, the clustered hyperplanes generally do not coincide with the decision boundary of the total neural network for the part of the training set that is classified by that cluster. As a result of this, the number of training examples that are located at the wrong side of the clustered hyperplanes is non-minimum.

It can be shown that the deviation from the situation depicted in (9.12) increases with decreasing magnitude of the bias input of the neurons in the first layer[1]. It follows that

$$\Delta|\underline{\Delta W}^{\varepsilon}| = \vartheta(\alpha_1^{bias}) \quad \textit{with} \quad \frac{\partial\vartheta(\alpha_1^{bias})}{\partial\alpha_1^{bias}} < 0 \tag{9.13}$$

[1] This can be explained as follows. For a *large* bias input of neurons in the first layer, in the beginning of training the hyperplanes of these neurons adapt fast towards the desired decision boundary. The weights of the neuron in the second layer then adapt to approximatey satisfy (9.12). For smaller bias inputs, the hyperplanes of the neurons in the first layer adapt more slowly towards the decision boundary, which is compensated for by another adaptation of the weights of the neuron in the second layer. As a result the distance between the hyperplanes of neurons in the first layer and the decision boundary of the network increases with decreasing bias input of neurons in the first layer.

Conclusion
It can be concluded from (9.9) and (9.13) that the value of β_1^h for which the summed terms between brackets becomes positive increases monotonously with α_1^{bias}: With the definition of β_{crit} in section 8.5:

$$\beta_{crit} = \Pi(\alpha_1^{bias}) \quad with \quad \frac{\partial \Pi(\alpha_1^{bias})}{\partial \alpha_1^{bias}} > 0 \tag{9.14}$$

The time spent in temporary minima depends on the number of training examples required to 'break' the clustered hyperplanes into multiple clusters. It follows from (9.9), (9.13) and (9.14) that both the "tightness" of the clustering increases (i.e. the minimum φ or *dist* is smaller) with α_1^{bias} *and* that the number of training examples required to reach β_{crit} increases with α_1^{bias}. Denoting the 'sticking' time in temporary minima as τ_{tm},

$$\tau_{tm} = \Theta(\alpha_1^{bias}) \qquad with \qquad \frac{\partial \Theta(\alpha_1^{bias})}{\partial \alpha_1^{bias}} > 0 \tag{9.15}$$

Note that in (9.15) all parameters, excluding the bias input of the neurons in the first layer, are assumed to be fixed. These fixed parameters include among others the training set, the way in which the hyperplanes are clustered and the bias signal applied to the neuron in the second layer.

9.2.4 Overall conclusions

It was shown in chapter 8 that the learning behavior of two-layer neural networks can be subdivided into (after initial training) iteratively:

- learning outside temporary minima and
- learning inside temporary minima.

Learning outside temporary minima leads towards the encounter of a minimum. This minimum may be local, global or temporary. For the learning outside temporary minima, the required training time increases with decreasing $\Delta\beta^h$. With relation (9.7) it can be concluded that to reach a minimum as fast as possible, the bias input signal for the neurons in the first layer should be large:

$$\textit{fast learning outside minima} \quad \Rightarrow \quad \textit{large } \alpha_1^{bias} \tag{9.16}$$

After reaching a minimum, there are two possibilities: the minimum is local (or global which is a special type of local minima) or is temporary. In the first case, learning is finished. In the second case the network should escape from the temporary minimum as fast as possible. It follows from (9.15) that fast escaping from temporary minimum requires a small bias input signal:

$$\textit{fast escaping temporary minima} \quad \Rightarrow \quad \textit{small } \alpha_1^{bias} \tag{9.17}$$

Hence, there is a trade-off between time required to reach temporary minima and training time required to escape from temporary minima. From simulations, it appears that a good compromise is to set the first-layer biases to

$$\alpha_{1,j}^{bias,opt} = | \underline{U}_{1,j}^{CG} - \alpha_{1,j}^{bias,opt} \underline{B}_1^{bias} | . \tag{9.18}$$

Note that the optimal bias signal of first-layer neurons is generally different for all neurons in the first layer. An illustration of the analyses and discussions in this section is presented in the next sub-section.

9.2.5 An illustration

An illustration of the dependency of both the training time required to each temporary minima and of the 'sticking' time in temporary minima on the first-layer neurons' biases is shown in figure 9.2. In this figure, simulation results are shown for 5 ratios between $\alpha_{1,n}^{bias}$ and α^{ATT}.

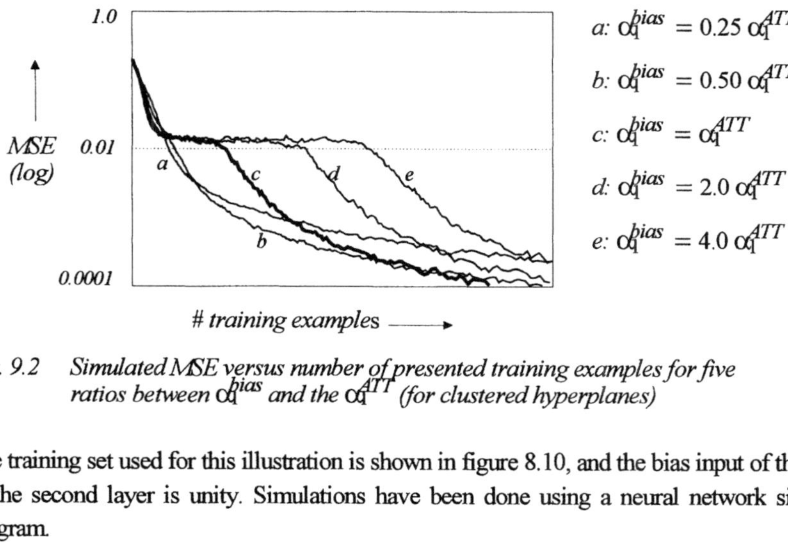

Fig. 9.2 Simulated MSE versus number of presented training examples for five ratios between α_1^{bias} and the α_1^{ATT} (for clustered hyperplanes)

The training set used for this illustration is shown in figure 8.10, and the bias input of the neuron in the second layer is unity. Simulations have been done using a neural network simulation program.

It follows directly from figure 9.2 that training time that is required to reach a temporary minimum is minimum for $\alpha_1^{bias} \gg \alpha_1^{ATT}$, which was indicated by (9.7). For smaller bias input for neurons in the first layer, the required training time to reach minima increases.

According to (9.15), the 'sticking' time in temporary minima is a monotonously decreasing function of the the first-layer bias signals. In the figure, this 'sticking' time is non-zero for (c), (d) and (e); while curves (a) and (b) do not significantly 'stick' in a temporary minima. The 'sticking' time versus bias input relation corresponds to (9.15).

9.3 Effect of the second layer's bias signal on learning

In the previous section of this chapter, the effect of the bias signal applied to first-layer neurons of a two-layer net was analyzed. It was shown that a large as well as a small bias signal for neurons in the first layer slows down learning significantly. A compromise between training time inside and outside temporary minima was given in (9.18).

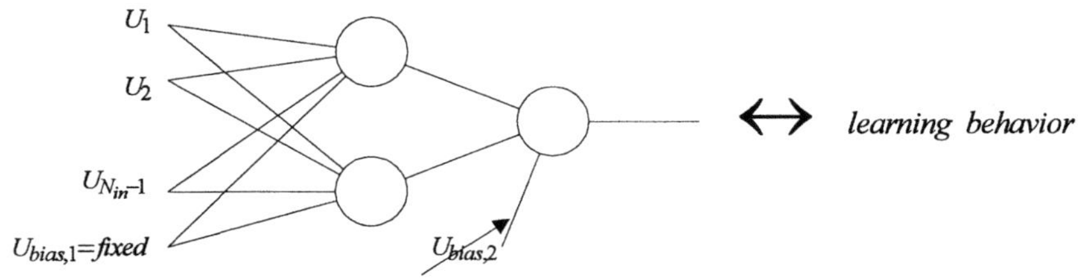

Fig. 9.3 The effect of the second-layer bias on the learning behavior is analyzed in this section

Similarly to the first-layer biases, it may be expected that the second-layer bias signal also has an effect on the required training time to reach a pre-specified level of performance. In section 9.3.1, a brief analysis of the effect of the magnitude of the second layer's bias input signal on the adaptation of $\underline{B}_{i,j}^{h}$ is presented. Section 9.3.2 discusses the relation between the second-layer's bias and the sticking time in temporary minima. The results of these analyses and discussions are summarized in section 9.3.3 and section 9.3.4 presents an illustrative example.

9.3.1 Second layer's bias input versus adaptation in the $\underline{B}^{h}$ direction

It has been derived in chapter 3 that the value of the bias signal applied to a specific neuron has a direct effect on the neuron's weight adaptation in the $\underline{B}_{k,j}^{h}$ direction. For small learning rates, the weight vectors of neurons in the first layer are approximately constant (or quasi-static) in a limited time interval. By decomposing $\alpha_{2,1}^{h}$ and $\underline{U}_{2,1}^{F}$ as

$$\alpha_{2,1}^{h} = \alpha_{2,1}^{h,\gamma} + \alpha_{2,1}^{ATT}$$

and

$$\underline{U}_{2,1}^{F} = \underline{U}_{2,1}^{F,\gamma} + \underline{U}_{2,1}^{F,const},$$

it can be derived that for a limited (training) time interval

$$\Delta\beta_{2,1}^{h} = \frac{\eta'}{P}\sum_{p}(D^{p}-Y^{p})f'(W_2\cdot\underline{U}_2^{p})\ \alpha_{2,1}^{h,\gamma,p} - \frac{\alpha_{2,1}^{ATT}\ \underline{U}_{2,1}^{F,const}\cdot\left(\Delta\overline{W_{2,1}^{\varepsilon}}\right)_{w\varepsilon}}{\alpha_2^{bias^2}+|\ \underline{U}_{2,1}^{F,const}\ |^2+\alpha_{2,1}^{ATT^2}} \tag{9.19}$$

$$\text{with } \eta' = \eta \left(\frac{\alpha_2^{bias^2} + |\underline{U}_{2,1}^{F,const}|^2}{\alpha_2^{bias^2} + \alpha_{2,1}^{ATT^2} + |\underline{U}_{2,1}^{F,const}|^2} \right)$$

It follows from (9.19) that for any point in training, the average adaptation of $\beta_{2,1}^h$ is a monotonously increasing function of $\alpha_{2,1}^{bias}$. Because the weights of $\underline{W}_{2,1}$ are included in the weight adaptation of first-layer neurons, a larger adaptation of $\beta_{2,1}^h$ results in an indirectly enlarged adaptation of the weight vectors of neurons in the first layer. However, because of the direct effect of $\alpha_{2,1}^{bias}$ on the adaptation of $\underline{W}_{2,1}$ and the indirect effect on the adaptations of $\underline{W}_{1,n}$, it can be concluded that throughout the training, the ratio between $\beta_{2,1}^h$ and $\beta_{1,j}^h$ increases with $\alpha_{2,1}^{bias}$:

$$\frac{\beta_{2,1}^h}{\beta_{1,j}^h} = \varsigma(\alpha_2^{bias}) \qquad with \qquad \frac{\partial \varsigma(\alpha_2^{bias})}{\partial \alpha_2^{bias}} > 0. \tag{9.20}$$

9.3.2 Relation between second layer's bias input and temporary minima

In this sub-section, the relation between the bias of the second-layer neurons and sticking time in temporary minima is analyzed and discussed. In this, we will follow the same line as in section 9.2.3.

During training a feed-forward neural net on a pattern classification task, the neurons in the last layer are usually activated in the full output range,

$$Y_{net} \in \langle f(-\infty), f(\infty) \rangle.$$

When training the neural network to perform a classification task, we assume that a double threshold method is used [1]. In this case, the network typically adapts on examples in-between the two thresholds[1]. Hence, the range of responses for which the network adapts is

$$Y_{net} \in \langle f(-\infty) + threshold \;,\; f(\infty) - threshold \rangle. \tag{9.21}$$

With 9.21) it follows that for examples that result in adaptation, the weighted input of the second-layer neuron satisfies

$$|\,\underline{W}_{2,1} \cdot \underline{U}_2\,| < f^{-1}\left(f(\infty) - threshold\right) \tag{9.22}$$

[1] Note that this assumes that there be no or at most very few training examples that result in an error $(D-Y)$ larger than $\left(f(\infty) - f(-\infty) - 2\, threshold\right)$, which assumption is in general valid.

For a cluster of m neurons in the first layer classifying (a part of) the training set, the hyperplanes are very close. This means that the response of all m first-layer neurons in a cluster on training examples is approximately equal. Denoting the index of the m neurons in the cluster as c with $c \in \{c_1, c_2, \ldots, c_m\}$, it follows from (9.5a) and (9.22) that for training examples that result in (a significant) adaptation of the weights of these clustered neurons

$$f\left(\alpha_{1,c}^{h,v}\,\beta_{1,c}^{h}\right) < \frac{1}{m\,(\underline{W}_{2,1})_c}\, f^{-1}\left(f(\infty) - threshold\right). \tag{9.23}$$

Note that only examples of the part of the training set that is (actively) classified by the cluster result in a significant weight adaptation for neurons in the cluster, which is due to the inclusion of the first derivative of the transfer function of these neurons in the weight update, see also (9.4).

Assuming that β_1^h is of the same order as $\beta_{2,1}^h$, it follows that for examples resulting in a significant adaptation, the neurons in the first layer are activated around the middle region. For sigmoid type transfer functions, (9.24) generally holds.

$$\Delta\underline{W}_{kj} \neq \underline{0} \qquad \Rightarrow \qquad f'''(\underline{W}_{1j} \cdot \underline{U}_1) < 0 \tag{9.24}$$

It follows from (8.26), (8.29) and (8.48) that the absolute value of the adaptation of $|\Delta\underline{W}^{\varepsilon}|$, $\Delta\beta_1^{bias}$ and β_1^h is a function of $\beta_{2,1}^h$. With (9.24), the $\beta_{2,1}^h$ merely acts as a scale factor for these three adaptations from which it follows that:

$$\frac{\Delta\beta_1^h}{\Delta|\Delta\underline{W}^{\varepsilon}|} = \Omega(\beta_2^h) \qquad with \qquad \frac{\partial\Omega(\beta_2^h)}{\partial\beta_2^h} \approx 0. \tag{9.25}$$

and

$$\frac{\Delta\beta_1^h}{\Delta\Delta\beta_1^{bias}} = \Xi(\beta_2^h) \qquad with \qquad \frac{\partial\Xi(\beta_2^h)}{\partial\beta_2^h} \approx 0. \tag{9.26}$$

With (9.1) and (9.25) it can be concluded that the value of β_{crit} for the rotation-based clustering mechanism, see (8.44), is by approximation independent of the value $\beta_{2,1}^{bias}$. Combining (9.2) and (9.26) results in the same conclusion for β_{crit} for the translation-based clustering mechanism. *It follows that the 'sticking' time in temporary minima is only effected by the difference between* β_1^h *and* β_{crit}, *due to different* $\beta_{2,1}^{bias}$, *at the encounter of the temporary minimum.*

During training outside temporary minima, the β_1^h of clustered neurons and β_2^h have values which are related. For a certain level of performance, and assuming correctly positioned hyperplanes, a certain gain is required by the whole network, which gain is distributed over both layers. It follows that for a certain performance:

$$\beta_1^h = \Psi(\beta_2^h) \qquad with \qquad \frac{\partial \Psi(\beta_2^h)}{\partial \beta_2^h} < 0. \tag{9.27}$$

With (9.20) it follows from (9.27) that the β_1^h at the encounter of a temporary minima decreases monotonically with β_2^{bias}. It follows from (9.1), (9.8) and (9.25) that the 'sticking' time in the encountered temporary minimum increases monotonically with β_2^{bias} for the rotation-based clustering mechanism. With (9.2), (9.8) and (9.26), the same can be derived for the translation-based mechanism:

$$\tau_{tm} = \zeta(\alpha_2^{bias}) \quad with \quad \frac{\partial \zeta(\alpha_2^{bias})}{\alpha_2^{bias}} > 0. \tag{9.28}$$

Note that if the β_1^h at the encounter of a temporary minimum is larger than β_{crit}, the 'sticking' time in the temporary minimum is zero. Consequently the temporary minimum may not be recognized from the *MSE*-training time curve.

9.3.3 Conclusions

The training time required to reach a specific level of performance of a two-layer neural network depends on the bias signal of the second-layer neuron. It is shown that the convergence outside temporary minima is maximum for large bias inputs because in that case both β_1^h and β_2^h have a large average adaptation.

However, 'sticking' time in temporary minima is large for large α_2^{bias}. This is due to the fact that the β_1^h at the encounter of a temporary minimum decreases with α_2^{bias} while β_{crit} is approximately independent of α_2^{bias}. On the other hand, for small α_2^{bias}, the convergence outside temporary minima is slow and 'sticking' time in temporary minima is short.

A good compromise between 'sticking' time and convergence outside temporary minima appears to be

$$\alpha_{2,1}^{bias,opt} = |\, \underline{U}_{2,1}^{CG} - \alpha_{2,1}^{bias,opt}\, \underline{B}_2^{bias} |\,. \tag{9.29}$$

Note that the optimal bias input for the *second* layer and the optimal bias input for neurons in the *first* layer are given by the same equations (not by the same values). In the following subsection an illustration of the conclusions of this section is presented.

9.3.4 An illustration

Figure 9.4 illustrates the results of this section. In figure 9.4, simulation results are shown for five different ratios between α_2^{bias} and α_2^{ATT}; the thick line corresponds to the situation in which $\alpha_2^{bias} = \alpha_2^{ATT}$. The training set used in these simulations is identical to the set used in section 9.2.5; the bias input of first-layer neurons were unity.

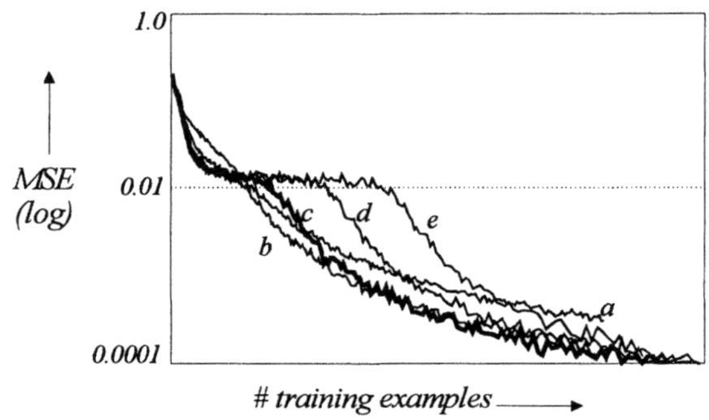

a: $\alpha_2^{bias} = 0.25\,\alpha_2^{ATT}$

b: $\alpha_2^{bias} = 0.50\,\alpha_2^{ATT}$

c: $\alpha_2^{bias} = \alpha_2^{ATT}$

d: $\alpha_2^{bias} = 2.0\,\alpha_2^{ATT}$

e: $\alpha_2^{bias} = 4.0\,\alpha_2^{ATT}$

Fig. 9.4 Simulated MSE versus number of presented training examples for five ratios between α_2^{bias} and the α_2^{ATT}

As argumented in section 9.3.1, the adaptation of the weight vectors in the $\underline{B}_{kj}^h$ direction increases with α_2^{bias}. In figure 9.4, this is illustrated by the convergence outside temporary minima which is faster for larger α_2^{bias}.
According to (9.28), the 'sticking' time in temporary minima decreases with α_2^{bias}. In figure 9.4, the 'sticking' time in temporary minima is approximately zero for cases (a) and (b) which is due to the fact that for these curves $\beta_1^h > \beta_{crit}$ at the encounter of the temporary minimum. For (c) to (e), the β_1^h at the encounter of a temporary minimum is lower than β_{crit} which results in significantly 'sticking' in temporary minima. The dependency of the 'sticking' time on the α_2^{bias} is large because the adaptation of β_1^h in temporary minima is relatively slow.

9.4 Large neural network: a problem and a solution

In large two-layer neural networks with one output, the training set S is usually subdivided into a large number of sub-sets S_i, each sub-set classified by a different cluster of (first-layer) hyperplanes. Note that a cluster may consist of only one hyperplane. In formula,

$$S = S_1 \cup S_2 \cup \ldots \cup S_n.$$

Assuming that each cluster of hyperplanes makes a significantly different classification than other clusters, the overlap between sub-sets $S_i \cap S_j$ is small for $i \neq j$. This means that the distance between training examples in sub-set S_i and the hyperplanes of neurons in the first layer in the jth cluster (with $i \neq j$) is generally large. It follows that the weighted input of neurons in the ith cluster on examples of the jth sub-set (with $i \neq j$) is generally large. Therefore, the response of these neurons satisfy:

$$Y_{1,i}\left(\underline{U}_1 \in S_j \mid i \neq j\right) \;\rightarrow\; f(\pm\infty) \tag{9.30}$$

During adaptation of weights on training examples out of sub-set S_i, the second-layer neuron virtually has a large bias input. This bias input vector consists of the trivial bias signal that is applied to this neuron *and* consists of the constant responses of neurons classifying other sub-sets. An illustration is given in figure 9.5.

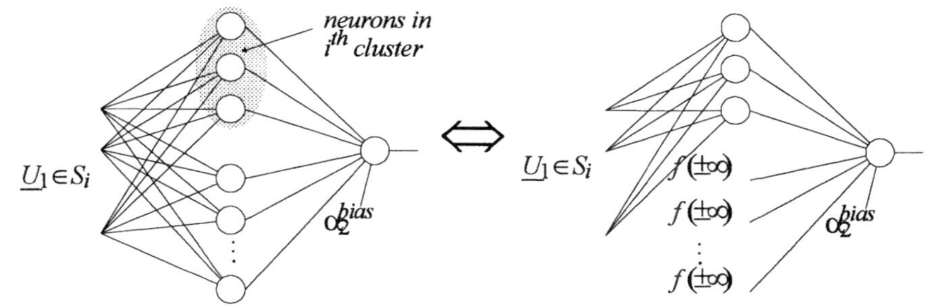

Fig. 9.5 Two layer neural network with a large first layer; the bias input of the neuron in the second layer is virtually larger than α_2^{bias}

For the example in figure 9.5, the virtual bias input vector applied to the neuron in the second layer, during adaptation on $\underline{U}_1 \in S_i$, is then (by approximation)

$$\alpha_{2,1}^{bias,virtual}\ \underline{B}_{2,1}^{bias,virtual} = \left[\ 0, 0, 0, f(\pm\infty), \ldots, f(\pm\infty), \alpha_2^{bias} \right]^T \tag{9.31}$$

Consequences for sticking time
For large single-output two-layer neural networks, there is a large number of first-layer neurons. Furthermore, the total training set S is typically subdivided into a large number of sub-sets. This means that $\alpha_{2,1}^{bias,virtual}$ is generally much large than the actual $\alpha_{2,1}^{bias}$:

$$\text{'}S\text{'}\ \textit{significantly subdivided} \quad \Rightarrow \quad \alpha_{2,1}^{bias,virtual} \gg \alpha_{2,1}^{bias} \tag{9.32}$$

It has been derived in section 9.3 that both the sticking time in temporary minima increases with the bias signal applied to the neuron in the second layer *and* that the convergence towards any minimum speeds up with the bias applied to the second-layer's neuron. With the conclusions of section 9.3, it follows that a neural network with a large first layer (and trained with back-propagation) generally converges fast towards minima, but 'sticks' for a relatively long (training) time in temporary minima.

Consequences for encountering local minima
In the beginning of training, the hyperplanes of neurons in the first layer cluster. This initial cluster may be positioned outside the training set. It is now possible that after breaking of the cluster, a number of hyperplanes of neurons in the first layer stay outside the training set and hence have a constant response [9].
In this case, the neuron in the second layer uses these outputs as "real" bias input signals: these output signals always serve to enlarge $\alpha_{2,1}^{bias,virtual}$. It is straight forward to show that neurons that are used to only bias neuron(s) in the next layer are not likely to get actively used to classify the training set. It follows that in this case the bias applied to the neuron in the second layer is virtually increased *and* that some neurons in the first layer are redundant (with the bias applied to the second-layer's neuron). If the number of remaining non-redundant neurons in the first layer is too small to correctly classify the training set, the neural network gets stuck in a local minimum.

A solution

To solve the problem of using virtually constant outputs of neurons in the first layer as an extra bias input of the neurons in the second layer, the back-propagation learning algorithm can be modified. We propose to only adapt the weight $(\underline{W}_{2,1})_i$ corresponding to a neuron in the first layer if the weight adaptation of $\underline{W}_{1,i}$ is not-negligibly small. In formula,

$$\Delta(\underline{W}_{2,1}^p)_i = \eta\left(D^p - Y^p\right) f'\left(\underline{W}_2 \cdot \underline{U}_2^p\right) \left(\underline{U}_{2,1}\right)_i^p \, \omega\left(f'\left(\underline{W}_{1,i} \cdot \underline{U}^p\right)\right), \tag{9.33}$$

where

$$\omega(x) = \begin{cases} 1 & x \geq threshold \\ 0 & x < threshold \end{cases}.$$

With (9.30) it can be concluded that the bias vector applied to the second-layer neuron on adaptation on $\underline{U}_1 \in S_i$, is by approximation $\alpha_2^{bias} \underline{B}_2^{bias}$. To ensure that the initial clustering process is not significantly changed, it may be useful to apply this modification of the weight adaptation rule only *after* initial training: the factor $\omega\left(f'\left(\underline{W}_{1,i} \cdot \underline{U}_i\right)\right)$ may degrade initial learning.

An illustration

In this section, an illustration of the effect of using a "large" first layer on the learning behavior is given. The training set used in this example is presented in figure 9.7a and 9.7b; the neural network must classify the inner square as "class 1" and the dotted outer region as "class 2". A (3,8,1) neural network was used in the simulations using this training set, i.e. a two-layer neural network with a three dimensional input vector (including the bias input), having eight first-layer neurons and with one neuron in the second layer was used. For comparison reasons, the weight adaptation is performed by both the back-propagation algorithm, and has been done using back-propagation with the modification in (9.33).

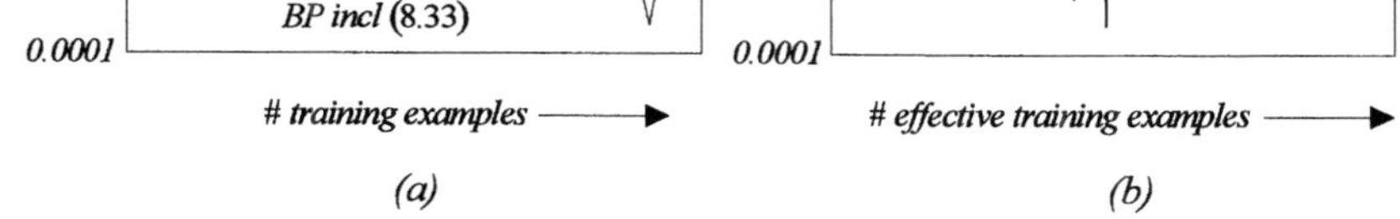

Fig. 9.6 (a) the MSE versus number of presented training examples for the original and modified back-propagation algorithm, and (b) the MSE versus effective training examples for both algorithms

Figure 9.6a shows the simulated MSE versus the number of presented training examples for both the original back-propagation algorithm (upper curve) and for the back-propagation algorithm with the modification (9.33). At the end of training, the network trained with the original back-propagation algorithm classified 72% of the training examples correctly, while the network that uses the modified algorithm classifies 94% correctly. These simulations illustrate clearly that the modified algorithm out-performs the original back-prop algorithm.
The difference between the two algorithm becomes even more evident if the MSE is plotted as a function of the number of training examples from which the neural nets learn. This is done because at the end of training, the neural network with the back-propagation algorithm adapts on 28% of the presented examples while the net with the modified algorithm adapts on only 6% of the presented training examples. In figure 9.4b, the training examples that result in weight adaptation have been denoted as *effective training examples*. By eliminating training examples that are correctly classified from the training set, as for example done by Romaniuk [8], the training time is reduced because the network now adapts on (approximately) all examples in the training set instead of adapting on only the (generally small part of the training set that consists of) examples that are not yet correctly classified.

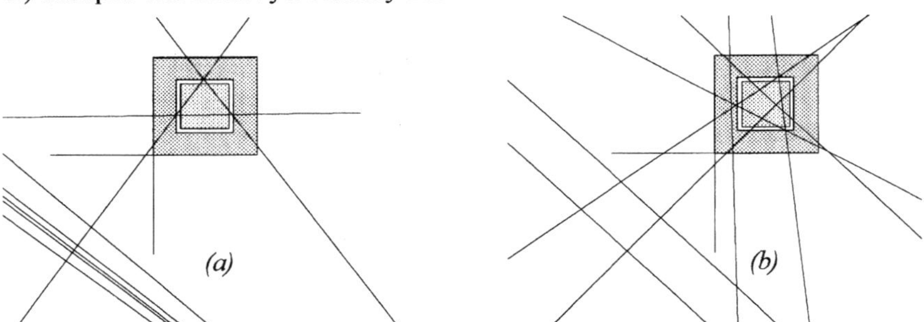

Fig. 9.7 Training set and the eventual hyperplane (of neurons in 1st layer) configuration for (a) the back-propagation algorithm (b) the modified back-prop algorithm

The curves in figure 9.6a and figure 9.6b that correspond to the original back-propagation algorithm indicate that it is likely that that neural net is stuck in a local minimum[1]. To illustrate this, figure 9.7a presents the position of the eight hyperplanes corresponding to the first-layer neurons after the training. Three of these hyperplanes are actively used to classify the training set, which results in a classification performance of 72% percent. The other five hyperplanes are positioned (relatively far) outside the training set and have an approximately constant response.

[1] This minimum might be a temporary minimum, but extended simulations show that the encountered minimum is very likely to be a local minimum; no significant decrease of the MSE is found for a ten times longer training nor do clustered hyperplanes (see figure 9.5a) tend to break into multiple clusters.

Figure 9.7b shows the position of the eight hyperplanes that correspond to the neurons in the first layer after training using the modified back-propagation algorithm. Six out of the eight neurons in the first layer are now actively used to classify the training set, which results in a classification performance of 94% (increasing during continued training).

Conclusions

With the (original) back-propagation algorithm, training a neural net with a first layer consisting of a large number of neurons results in a number of problems. These problems include long sticking times in temporary minima, having redundant neurons in the network, and the resulting increased probability to encounter a local minimum. A simple modification of the weight adaptation rule for the weights of the neuron(s) in the second layer reduces these problems significantly.

9.5 Unipolar input signals

It has been argumented in chapter 4 that it is advantageous to use two-quadrant multipliers instead of four-quadrant multipliers for the synapses in hardware implementations of neural networks. The advantages of the two-quadrant multipliers over the (equally fast) four-quadrant multipliers are the usually smaller required chip area and the lower power consumption. To be able to apply two-quadrant multipliers in the first layer of a neural net, the training set may need to be shifted in input space to obtain unipolar input signals. For single-layer neural networks, shifting the training set in input space has a direct effect on the weight adaptation. It has been shown in chapter 4 that the training time can be minimized by proper adjustment of either the bias input signal or the learning rate of only this bias input signal.

In this section, the effect on training time of shifting the training set in input space, such that unipolar input signals are obtained for a *two-layer* neural network, is discussed.

Discussion

In section 9.2 and 9.3, it has been demonstrated that the learning behavior of two-layer feed-forward neural nets can be optimized by properly setting of the bias signals of all neurons. The optimal value for the bias signal as presented in sections 9.2 and 9.2 are a compromise between fast convergence outside temporary minima and fast escaping from temporary minima:

$$\alpha_{i,j}^{bias} \gg \sqrt{\alpha_{i,j}^{ATT^2} + |\underline{U}_{i,j}^{F,constant}|^2} \quad \rightarrow \textit{fast reaching minima.}$$

and

$$\alpha_{i,j}^{bias} \ll \sqrt{\alpha_{i,j}^{ATT^2} + |\underline{U}_{i,j}^{F,constant}|^2} \quad \rightarrow \textit{fast 'escaping' temporary minima.}$$

It was shown in sections 9.2 and 9.3 that a good compromise between short 'sticking' time and fast convergence outside temporary minima obtained for:

$$\alpha_{i,j}^{bias,opt} = | \underline{U}_{i,j}^{CG} - \alpha_{i,j}^{bias,opt} \underline{B}_2^{bias} | . \tag{9.34}$$

When shifting the training set in input space in such a way that unipolar input signals for the neural network are obtained, the norm of the vector pointing towards the centre-of-gravities corresponding to clustered hyperplanes of neurons in the first layer generally increases. This means that both the convergence inside and outside temporary minima is changed due to a translation of the training set in input space (see sections 9.2 and 9.3). The training time can again be minimized by setting the biases according to (9.34).

9.6 References

[1] E.D.Sontag and H.J.Sussmann, "Backpropagation Separates when Perceptrons Do", in Proc. IJCNN, Washington D.C., vol. 1, pp. 639-642, 1989

[2] D.E.Rumelhart, G.E.Hinton and R.J.Williams, "Learning internal representations by error propagation", in *Parallel Distributed Processing: Exploration in the microstructure of cognition* (Vol. 1), Cambridge MA: MIT Press, 1986

[3] S.J.Hanson and D.J.Burr, "Minkowski-r Back-Propagation:Learning in Connectionist Models with Non-Euclidian Error Signals", in *Neural Information Processing Systems*, ed. D.Z.Anderson, AIP: New York, pp. 347-357, 1988

[4] K.Matsuoka and J.Yi, "Backpropagation Based on the Logarithmic Error Function and Elimination of Local Minima", in Proc. IJCNN, Singapore, vol. 2, pp. 1117-1122, 1991

[5] M.J.J.Holt and S.Semnani, "Convergence of Back-Propagation in Neural Netwokrs using a Log-Likelyhood Cost Function", Electronic Letters, vol. 26, no. 23, pp. 1964-1965, 1990

[6] J.R.Movellan, "Error Functions to Improve Noise Resistance and Generalization in Backpropagation Networks", in Proc. IJCNN, Washington DC., vol. 1, pp. 557-560, 1990

[7] S.D.Wang and C.H.Hsu, "A Self Growing Learning Algorithm for Determining the Appropriate Number of Hidden Units", in Proc. IJCNN, Singapore, vol. 2, pp. 1098-1104, 1991

[8] S.G.Romaniuk and L.O.Hall, "Dynamic Neural Networks with the use of Divide and Conquer", in Proc. IJCNN, Baltimore, vol.1, pp. 658-663, 1992

[9] L.F.A.Wessels and E.Barnard, "Avoiding False Local Minima by Proper Initialization of Connections", IEEE tr. Neural Networks, vol. 3, pp. 899-905, 1992

Chapter 10

Cost Functions for Two-Layer Neural Networks

10.1 Introduction

With the original formulation of the back-propagation algorithm, a Mean Squared Error cost function is minimized (see chapters 1, 3 and 8). The classification (and mapping) capabilities of feed-forward neural networks trained with back-propagation were discussed in section 1.2. It was shown that:

- two-layer feed-forward neural networks are capable of approximating any function arbitrarily well (theorem 1.2) and
- feed-forward neural networks applied to classifying patterns, and trained with an algorithm that minimize the MSE criterion, implements a Bayesian estimator on a posteriori class probability (theorem 1.6)

The learning behavior of single-layer and two-layer feed-forward neural networks trained with back-propagation were analyzed in chapters 3 to 9. In these chapters, it was assumed that the original formulation of back-propagation was used i.e. it was assumed that the MSE cost function was minimized. It is however not necessary in gradient descent optimization methods such as back-propagation to minimize this MSE cost function.
This chapter deals with the learning behavior of feed-forward neural networks that have been trained using a cost function which is the mean of a certain rational power of the error. It will be shown that for fast convergence towards temporary minima and fast convergence out of temporary minima conflicting demands are posed upon the cost function. This can be used to synthesize an algorithm that switches between cost functions whenever this would speed up learning.

10.2 Discussion of "Minkowski-r back-propagation"

The back-propagation algorithm as defined by Rumelhart et al. [9] uses the Mean Squared Error cost function which is to be minimized during training. This cost function is also used in most back-propagation derivatives. In 1988, Hanson and Burr introduced the so-called Minkowski-r back-propagation algorithm [3]; this algorithm is basically back-propagation, with a modified energy function:

$$E_r = \frac{1}{r \cdot P} \sum_{p=1}^{P} \left| D^p - Y^p \right|^r \tag{10.1}$$

In [3], the training time required to reach a specific level of performance is reported to increase roughly with the parameter r. However in noisy environments, a small r is reported to improve convergence to correct classification of the training set. Other papers on modifying the energy function [4]-[6] report on faster convergence and higher noise-immunity. Because the analysis of the effect of (10.1) can be extended towards the analysis of other energy functions, this section is restricted to the energy function in the Minkowski-r back propagation algorithm.
In this section, results from the analyses in chapter 3 and 9 will be used to analyze the effect on training time of applying an energy function which is the summed r^{th} power of the error. A trivial result of using another energy function is that another type of error is minimized. This means that for example for E_1 the difference D–Y is minimized.

10.2.1 Making an "initial guess"

It has been shown in chapter 8 that in the beginning of training, two-layer neural nets attempt to separate the training set linearily. If this is detected to be impossible, the (then) clustered hyperplanes of the first-layer neurons that are actively used for the classification "break" into multiple clusters. It is shown in chapter 8 that this subdivision of the training set occurs only if this subdivision results in an improvement of the classification. In other words, the neural network makes at first an "initial guess" to classify the total training set using virtually one hyperplane, and then iteratively subdivides the total training set into manageable sub-sets in order to maximize the performance on the total training set.
The "initial guess" is important in training because it serves as the point from which the neural network starts to iteratively improve its classification. It can be shown that for training sets for which the "initial guess" is relatively good, i.e. for training sets for which the eventual classification boundary can be relatively good approximated by a single hyperplane, the order of the energy function does not significantly change the "initial guess".

However, for training set for which the optimal decision boundary cannot be approximated by one single hyperplane (see for example the training set in figure 9.7) the order r of the energy function E_r has a direct effect on the "initial guess". For the training set in figure 9.7, a first order energy function would result in positioning the hyperplane of all neurons in the first layer far outside the training set in order to classify most of the training examples correctly. For this "initial guess", the response on any training example is identical (either "low" or "high", depending on which class contains most examples). If this situation is encountered, the network is stuck in a local minimum. For these "complicated" training sets, higher order energy functions result in a better "initial guess".
The training set used for the illustrations in this section is shown in figure 8.7. For this relatively simple training set, the optimal decision boundary can be approximated by one single hyperplane.

10.2.2 Analysis of the training time required to reach minima

Outside temporary minima, the training time is dependent on the adaptation of any $\beta^h_{i,j}$ in the neural network. To be able to compare the training time for different r, the learning rate η is adjusted in such a way that the adaptation of weight vectors for small weights (as defined in section 8.3) is identical. With E_2 as a reference, it follows that

$$\eta_r = \eta_2 \; | D^p - f(0) |^{2-r} \qquad (10.2)$$

where:

- η_r is the learning rate for E_r
- D^p is the desired (target) response for an example (assuming double threshold targets that are symmetrical with respect to $f(0)$)
- $f(0)$ is the response of the neurons for zero weighted input

It has been shown in section 6.5 that for single-layer feed-forward neural networks, separable training examples tend to *increase* the $\beta^h_{i,j}$ while non-separable examples tend to *decrease* any $\beta^h_{i,j}$ (assuming the double-threshold variant of back-propagation [10]). It is straight-forward that the same holds for any neuron in a two-layer neural network, when compensating for negative signs in $\underline{B}^h_{2,1}$, see section 8.5). In formula:

$$\begin{cases} \underline{U}^{lin\_sep} & \Rightarrow \; \overline{|\Delta\beta^h_{i,j}|} \geq 0 \\ \underline{U}^{non\_lin\_sep} & \Rightarrow \; \overline{|\Delta\beta^h_{i,j}|} < 0 \end{cases} \qquad (10.3)$$

During training, the error $(D-Y)$ decreases monotonically during training for separable examples and increases for non-separable examples. Furthermore, the error $(D-Y)$ for non-separable examples is large than for separable examples:

$$(D-Y)_{sep} < (D-Y)_{non-sep} \tag{10.4}$$

with

$$\frac{\partial (D^p - Y^p)_{sep}}{\partial t} \leq 0 \quad and \quad \frac{\partial (D^p - Y^p)_{non\ sep}}{\partial t} > 0 .$$

Noting that the function x^r is strongly increasing with the exponent r, the following properties for separable and non-separable examples can readily be concluded:

- the average (positive) adaptation of $|\beta_{i,j}^h|$ decreases significantly with r for separable examples
- the average (negative) adaptation of $|\beta_{i,j}^h|$ decreases significantly with r for non-separable training examples

As the training time to reach a minimum (local, global or temporary) decreases with the speed of adaptation of all $|\beta_{i,j}^h|$, the *training time required to reach a minimum increases with r.*

10.2.3 Analysis of 'sticking' time in temporary minima

Equations (8.35) and (8.51) show that adaptation on training examples that are located at the "wrong" side of clustered hyperplanes (from a classification point of view) tend to 'break' the cluster. Other examples tend to further cluster the hyperplanes. It follows from the analysis in sections 8.4 to 8.6 that (for the E_2 energy function) the increment of the angle between hyperplanes in a cluster is a monotonically increasing function of the error $(D-Y)$. In formula

$$\Delta\varphi^p = \iota(|D-Y|) \quad with \quad \frac{\partial \iota(|D-Y|)}{\partial |D-Y|} > 0 \tag{10.5a}$$

and

$$\Delta dist^p = \kappa(|D-Y|) \quad with \quad \frac{\partial \kappa(|D-Y|)}{\partial |D-Y|} > 0. \tag{10.5b}$$

The 'sticking' time in temporary minima is among others determined by:

- the minimum angle φ between clustered hyperplanes and by the adaptation of β_1^h and $\underline{\Delta W}^\varepsilon$ in this temporary minimum, for the rotation-based clustering-de-clustering mechanism
- the minimum distance *dist* between clustered hyperplanes, the adaptations of β_1^h and of $\Delta\beta_1^{bias}$ in the minimum for the translation-based mechanism

Other factors affecting the 'sticking' time in temporary minima include the training set and the actual clustering; these parameters are however assumed to be fixed for comparison reasons. The minimum angle or distance between clustered hyperplanes corresponds to the φ or *dist* at the point at which the clustered hyperplanes start to de-cluster. In (8.44) and (8.53) this point has been denoted as τ_{crit}.

In the so-called Minkowski-r back-propagation algorithm, the weight adaptation scales with the derivative of the cost-function. Assuming identical learning rates η, the adaptation on any example is therefore increased by a factor $|D-Y|^{r-2}$ with respect to normal back-propagation (which uses E_2). With the inclusion of (10.2) it follows that for Minkowski-r back-propagation,

$$\Delta\varphi_r^p = \theta(|D-Y|, r) \tag{10.6a}$$

with

$$\frac{\partial\theta(|D-Y|, r)}{\partial\,|D-Y|} > 0 \quad and \quad \frac{\partial\theta(|D-Y|, r)}{\partial\, r}\cdot\left(\frac{\partial |D-Y|^{r-1}}{\partial r}\right)^{-1} > 0. \tag{10.6b}$$

The same holds for the translation-based breaking mechanism:

$$\Delta dist_r^p = \rho(|D-Y|, r) \tag{10.7a}$$

with

$$\frac{\partial\rho(|D-Y|, r)}{\partial\,|D-Y|} > 0 \quad and \quad \frac{\partial\rho(|D-Y|, r)}{\partial\, r}\cdot\left(\frac{\partial |D-Y|^{r-1}}{\partial r}\right)^{-1} > 0. \tag{10.7b}$$

Note that in the second expression in (10.6b) and (10.7b), the derivative of the $(r-1)^{th}$ power of the error must be included to compensate for the change in average weight adaptation due to changing r, i.e. to include (10.2). It can be concluded that *the quotient of 'sticking' time in temporary minima and training time required to reach a temporary minimum, decreases with r.*

10.2.4 An illustration

Figure 10.1 shows the simulated *MSE* versus number-of-training-examples curves for four different *r*. The training set used in the simulation is non-linearly separable, but is can be classified correctly by the selected network size; the learning rates η were scaled according to (10.2).

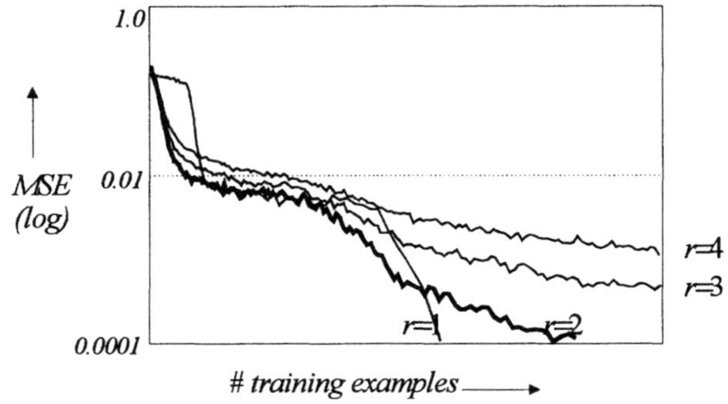

Fig. 10.1 MSE versus number of presented training examples for four values of r

Figure 10.1 shows clearly that the adaptation towards temporary (and global) minima decreases with *r*, as explained in section 10.2.4: the 'sticking' time in temporary minima, *related to the time required to reach the minimum*, decreases with *r*.

10.3 Switching cost functions

The advantage of using cost-functions with a high *r* is the relatively short 'sticking' time in temporary minima. However, a major disadvantage is the slower convergence to any minimum (temporary, local or global). A way to solve the either slow convergence towards minima or the relatively long 'sticking' time in temporary minima is to use two cost-functions. One cost-function which ensures fast convergence towards minima must then be used outside temporary minima, and the other cost-function must be used in temporary minima for fast 'escaping' from the minimum. It may be clear that the first cost-function has a low *r* (see 10.23), and that the second one has a high *r*.

To decide which cost-function must be used, information is needed on whether (a part of) the neural network is 'stuck' in a temporary minimum or not. An estimation on whether the network is stuck or not can be made in several ways.

An estimation on getting stuck

A fist estimation on getting stuck in a minimum can be obtained by using the mean squared error over only training examples that result in non-zero weight adaptation. The examples that result in weight adaptation are denoted as *effective examples*, and hence the signal is denoted as the Mean Squared Effective Error signal (MSEE). In formula:

$$MSEE = \frac{1}{P_{eff}} \sum_{peff=1}^{P_{eff}} \left(D^{peff} - Y^{peff}\right)^2 \tag{10.8}$$

It can be derived that this MSEE signal will have a minimum for a non-minimum MSE if and only if the training set contains misclassified examples (lying on the 'wrong' side of the classification boundary) and if during the continuing decrement of the MSE the classification performance on the separable training examples improves [1]. As this latter condition is usually satisfied in training neural networks, it follows that the MSEE can be used to detect the encountering of minima. This minimum can be either global, local or temporary:

$$min(MSEE) \Leftrightarrow encounter\ of\ minimum \tag{10.9}$$

Escaping a temporary minimum is detected as a decrease of the MSE after encountering a minimum. With these criteria, an algorithm that switches between E_1 and E_4 has been constructed. Simulation results, using a non-linearly separable training set which is separable using two (clusters of) hyperplanes, are shown in figure 10.2.

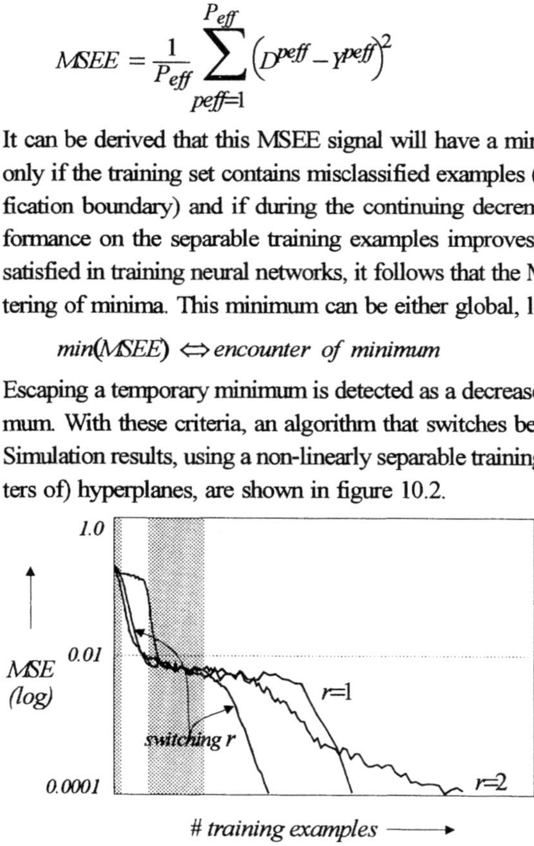

Fig. 10.2 MSE versus number of presented training examples for E_1, E_2 and an algorithm that switches between E_1 and E_4 (E_4 regions are dotted) using the MSEE

For the curve with switching r, the dotted regions in figure 10.2 correspond to using E_4 while the other regions correspond to usage of E_1. It follows from the simulation that the variable-r algorithm has a fast convergence towards minima, while the 'sticking' time in temporary minima is relatively short. However, this 'sticking' time in the temporary minimum is somewhat longer than might be expected from figure 10.1 because of switching between the two energy functions which causes some type of 'settling-time' delay for the network. The advantage of switching between energy functions increases with the sticking time in temporary minima.

Some other methods
Another method to detect possibly encountering or sticking in minima is using the relative decrease of the MSE during training [2],[8],[12]. For large relative decrements of the MSE, the network is not stuck in a minimum, and for small relative decrements of the MSE the neural network is likely to get stuck in a minimum. It has been shown that to escape temporary minima cost-functions with large r are better while convergence outside temporary minima is optimum for small r.

In [1] an algorithm that uses the relative decrement of the MSE to continuously vary the cost-function is presented. The order r of the cost-function to be used in the neural network is in this algorithm limited to a certain range $r \in [r_{min}, r_{max}]$. The proposed equation describing the order r is

$$r = \begin{cases} r_{min} & \mathit{for}\ r' < r_{min} \\ r_{max} & \mathit{for}\ r' > r_{max} \\ r' & \mathit{else} \end{cases} \tag{10.10}$$

with

$$r' = E\left(r_{max} + \Delta r - \xi \frac{\Delta MSE}{MSE}\right)$$

In this equation, the $E(\cdot)$ operator calculates a short term average (or moving average), and the ξ parameter is used to utilize the total range for r. To maintain a acceptable speed of convergence for higher order cost-functions, (10.2) is included to continuously scale the learning rate η.
Simulations show that both of these algorithm have performances which are comparable with the switching-cost-function algorithm presented earlier in this section.

10.4 Classification performances using non-MSE cost-function

It is well known from literature, see section 1.2, that a feed-forward neural network trained with a gradient descent method that minimized the MSE cost-function implements a Bayesian a posteriori estimator on class probability. It is however shown by a number of authors that when using other cost-functions, the classification performance after training is comparable with that of neural classifiers trained to minimize the MSE [7],[11].

Reported differences (and the goal) between gradient descent algorithms that minimize the MSE and algorithms that minimize other types of cost-functions include faster convergence during training, better generalization of the learned function towards vectors that were not in the training set, higher noise immunity, and higher fault tolerance of the neural network.

10.5 Conclusions

The analyses in section 10.3 show that the energy function used during training of a feed-forward neural network directly affects the learning behavior. It is shown that low-order cost-functions result in a fast convergence outside temporary minima but longer 'sticking' in these temporary minima. Higher order cost-functions result in a slower convergence outside the minima and in a short sticking in temporary minima.

Using *low order* cost-functions, problems may arise with training sets for which the optimal classification boundary cannot be approximated by one single hyperplane. In this case, the "initial guess" of the neural network may be very bad which results in an increased probability on the encounter of a local minima.

A few modifications of the back-propagation algorithm illustrate that when either switching between energy functions or when using variable energy functions, the convergence inside *and* outside temporary minima is fast.

10.6 References

[1] A.J.Annema, "Analysis, Modelling and Implementation of Analog Integrated Neural Networks", Ph.D. thesis, University of Twente, The Netherlands, 1994

[2] I.Gloeckner, "Monotonic Incremation of Backpropagation Networks", in Proc. ICANN 93, Amsterdam, pp. 498, 1993

[3] S.J.Hanson and D.J.Burr, “Minkowski-r Back-Propagation:Learning in Connectionist Models with Non-Euclidian Error Signals”, in *Neural Information Processing Systems*, ed. D.Z.Anderson, AIP: New York, pp. 347-357, 1988

[4] M.J.J.Holt and S.Semnani, “Convergence of Back-Propagation in Neural Netwokrs using a Log-Likelyhood Cost Function”, Electronic Letters, vol. 26, no. 23, pp. 1964-1965, 1990

[5] K.Matsuoka and J.Yi, “Backpropagation Based on the Logarithmic Error Function and Elimination of Local Minima”, in Proc. IJCNN, Singapore, vol. 2, pp. 1117-1122, 1991

[6] J.R.Movellan, “Error Functions to Improve Noise Resistance and Generalization in Backpropagation Networks”, in Proc. IJCNN, Washington DC., vol. 1, pp. 557-560, 1990

[7] M.D.Richard and R.P.Lippmann, "Neural Network Classifiers Estimate Bayesian a posteriori Probabilities", Neural Computation, vol. 3, pp. 461-483, 1991

[8] S.G.Romaniuk and L.O.Hall, “Dynamic Neural Networks with the use of Divide and Conquer”, in Proc. IJCNN, Baltimore, vol.1, pp. 658-663, 1992

[9] D.E.Rumelhart, G.E.Hinton and R.J.Williams, “Learning internal representations by error propagation”, in *Parallel Distributed Processing: Exploration in the microstructure of cognition* (Vol. 1), Cambridge MA: MIT Press, 1986

[10] E.D.Sontag and H.J.Sussmann, “Backpropagation Separates when Perceptrons Do”, in Proc. IJCNN, Washington D.C., vol. 1, pp. 639-642, 1989

[11] B.A.Telfer and H.H.Szu, "Energy Functions for Minimizing Misclassification Error With Minimum-Complexity Networks", Neural Networks, vol. 7, pp. 809-818, 1994

[12] S.D.Wang and C.H.Hsu, “A Self Growing Learning Algorithm for Determining the Appropriate Number of Hidden Units”, in Proc. IJCNN, Singapore, vol. 2, pp. 1098-1104, 1991

Chapter 11

Some issues for f'(x)

11.1 Introduction

In feed-forward neural networks, the response of neurons is a non-linear function of the weighted input of that neuron. This non-linear activation function is typically a saturating function for large weighted inputs. Widely used activation functions include the piece-wise linear function and sigmoid functions.

In the learning phase of a feed-forward neural network using back-propagation (or a derivative), the weight are adapted using a gradient descent method as to minimize a predefined cost-function (usually the MSE cost-function):

$$\Delta \underline{W}_{i,j} = -\eta \frac{\partial E}{\partial \underline{W}_{i,j}}$$

In the frequently used "on-line" approximation, the weights are adapted for each example:

$$\Delta \underline{W}_{i,j} = -\eta \frac{\partial (D^p - Y^p)^2}{\partial \underline{W}_{i,j}} .$$

As the response of the neural network includes the non-linear activation function, it follows from the chain-rule that the first derivative of the response of the non-linear function with respect to its input is required for weight adaptation, see also (8.7) and (8.8). Section 11.2 describes demands on the non-linear function and its derivative for single-layer neural networks for learning neural networks. After this discussion for single-layer networks, the demands on the non-linear function with derivative are presented for two-layer neural networks in section 11.3. It will be shown that by the inclusion of a mismatch between the actual derivative and the implemented derivative, the learning behavior of the multi-layer network can be speeded up considerably. Note that in this case, no true gradient descent algorithm is implemented.

11.2 Demands on the activation function for single-layer nets

It was illustrated in chapter 3, a single-layer feed-forward neural network with N_1 neurons is equivalent to N_1 single neurons operating independently, but having the same dimension for the input space. The response of single neurons is typically a non-linear function of the weighted input of that neuron. In figure 11.1 a single neuron and a typical non-linear function (the sigmoid) are shown.

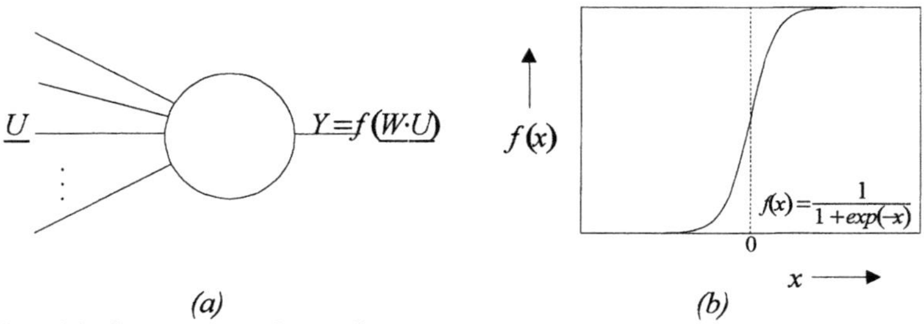

Fig. 11.1 (a) The response of a single neuron and
(b) a typical activation function for feed-forward neural nets: the sigmoid

For single-layer neural networks, the activation function has to satisfy a few demands, which are listed below. For every demand, a short discussion is given.

- Firstly, the activation function used in single-layer neural networks must be either monotonically increasing or monotonously decreasing with the input signal. For learning neural networks, this non-linear function must be differentiable in order to apply a gradient descent learning algorithm (such as back-propagation).

- For function approximation applications, the desired (or target) response for neuron is generally in a limited range. Therefore, it is generally sufficient for single-layer neural networks to have saturating activation functions. These non-linear saturating functions are often referred to as squashing-functions because of their property of squashing a large input range into a limited output range.

- When using single-layer neural networks as pattern classifiers, the input vectors $\underline{U}$ are typically classified into two classes. When using a saturating activation function, the response of the network on input vectors is ideally lower saturating or upper saturating, corresponding to logical "0" and logical "1" indicating the class in which the example is classified. Hence, the activation function needs to be saturating for pattern classification tasks.

- With gradient descent-based weight adaptation algorithms, the first derivative of the activation function with respect to its input must be included in weight adaptations. However, this inclusion of the derivative is not necessary to ensure full learning capabilities for single-layer feed-forward neural networks. Omission of the first derivative from the weight adaptation effectively corresponds to minimizing a non-*MSE* cost-function. For networks that have a sigmoid transfer function, it can be derived that the minimized cost-function is given by

$$E = D\ln(Y) - D\ln(D) + (1-D)\ln(1-Y) - (1-D)\ln(1-D) \tag{11.1}$$

Compared to the *MSE* cost-function minimized in back-propagation, the weight adaptation using the new cost-function larger for examples that result in either small or result in very large errors (examples that are almost classified correctly and examples that are classified incorrect). This is illustrated by the two curves in figure 11.2. One of the two curves corresponds to the adaptation of β^h with the back-propagation algorithm while the other curve corresponds to minimizing ().

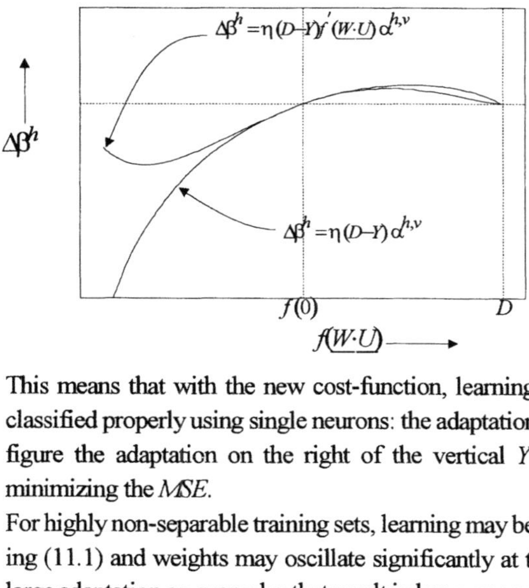

Fig. 11.2 The adaptation of β^h for two cost-functions (MSE and (11.1), by a gradient descent algorithm

This means that with the new cost-function, learning is faster for training sets that can be classified properly using single neurons: the adaptation on linearly separable examples (in the figure the adaptation on the right of the vertical $Y=f(0)$ line) is enlarged compared to minimizing the *MSE*.

For highly non-separable training sets, learning may be slower (see also (10.3) when minimizing (11.1) and weights may oscillate significantly at the end of training, which is due to the large adaptation on examples that result in large errors (on the very left side of the curve). This should not be a problem as for these sets, classification using single neurons will generally be bad.

11.3 Demands on the activation functions for two-layer nets

In single-output two-layer feed-forward neural networks, the network response is typically a non-linear function of the weighted input of the second-layer's neuron. The inputs of this neuron are provided by the first-layer neurons and a bias source. The output signals of the first-layer neurons are again a non-linear function of their weighted inputs.

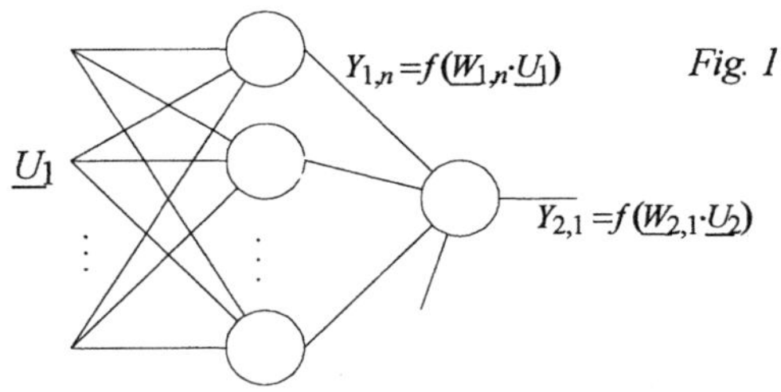

Fig. 11.3 The response of any neuron in a (here single-output two-layer) feed-forward neural network is generally a non-linear function of the weighted input

In two-layer feed-forward neural network (and in general multi-layer feed-forward networks), the activation function must satisfy more demands than the activation functions for single-layer nets. The demands for the first-layer activation function of two-layer network are discussed first. After this discussion, the demands on the second-layer's activation function presented.

Demands on the activation function for the first layer

In this part, some demands on the first-layer activation function are presented. It is assumed that this first layer is a part of a single-output two-layer neural network. The demands are however valid for every non-last layer in a general multi-layer feed-forward neural network. The demands are listed below:

- For any multi-layer feed-forward neural networks, the activation function of neurons that are not in the last layer must be non-linear (if the activation functions of layer N were linear, the Nth layer and the $(N+1)^{th}$ layer can be replaced by one layer, which can be illustrated by applying simple linear matrix algebra). Stronger demands on the activation function of first-layer neurons were discussed in section 1.2.3. It was shown by Leshno et al. (see section 1.2.3) that the *first-layer activation function needs to be a non-polynomial squashing function* (i.e. non-polynomial and both lower and upper saturating).

- As demonstrated in sections 8.4 and 8.6, the de-clustering of clustered hyperplanes requires the difference between the first derivatives of the clustered first-layer neurons. A trivial conclusion of the analyses in chapter 8 is therefore that *the derivative of the first-layer neurons must be included in the weight adaptation and that the derivative must be non-constant.*

- The previous demands on the activation function and the derivative do not require that the actual derivative of the activation function is used in the weight adaptations: there may be a mismatch between the actual derivative and the implemented "derivative".

In the next part of this section, some issues concerning matching between the actual derivative and the implemented "derivative" of first-layer activation functions are discussed. It will be shown that a deliberate mismatch can speed up convergence in multi-layer neural networks.

Matching properties for first-layer activation function derivatives

In the back-propagation algorithm, the exact derivative of the activation function, $\partial f(x)/\partial x$ or $f'(x)$, is required in the adaptation of any weight that directly or indirectly affects the response of that activation function. In this section, the effect on training of mismatch between the actual derivative $f'(x)$ and the implemented "derivative" $\tilde{f}'(x)$ of first-layer neurons in a two-layer feed-forward neural network is discussed. The specific type of mismatch analyzed in this section is a gain-mismatch:

$$\tilde{f}'(x, G) = f'(G \cdot x) \tag{11.2}$$

where

G is the gain mismatch parameter (for no mismatch $G=1$)

The type of mismatch between the actual and the used approximation of the derivative of the activation function is illustrated by figure 11.4. The dotted curves correspond to mismatch between $f'(x)$ and $\tilde{f}'(x)$.

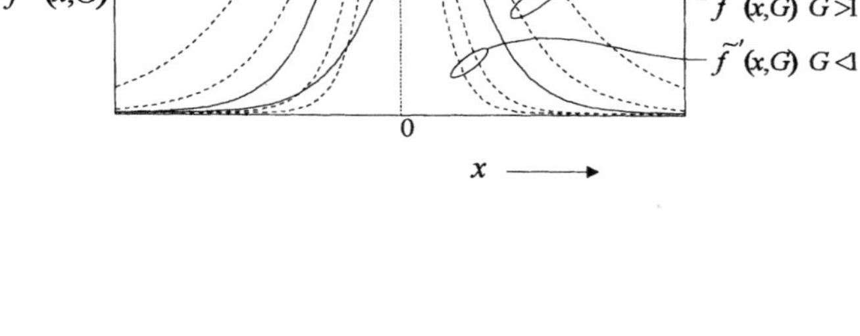

Fig. 11.4 Mismatch in the gain of the "derivative" of the activation function

It has been shown in chapter 8 that during training of a two-layer neural network, first redundancy between first-layer neurons is being build up and then is recursively abolished. Analyses in chapter 8 present two mechanisms that lead towards this type of behavior: the rotation-based mechanism (section 8.4) and the translation-based mechanism (section 8.6). These two mechanisms are basically described by the difference equations (8.31):

$$\Delta\varphi \approx \sin(\varphi)\cdot\left[\frac{\Delta|\Delta W^{\varepsilon}|}{|\Delta W^{\varepsilon}|} - \frac{\Delta\beta_1^h}{\beta_1^h}\right]$$

for the rotation-based mechanism and by (8.47):

$$\Delta dist \approx dist\cdot\left[\frac{\Delta|\beta_1^{bias}|}{|\beta_1^{bias}|} - \frac{\Delta\beta_1^h}{\beta_1^h}\right]$$

for the translation-based mechanism.

With major simplifications, the dynamic behavior can be approximated by the coupled difference equations given in sections 8.5 and 8.7. As mentioned in chapter 8, these sets of coupled equations are meant to get easily and relatively fast a good impression of the learning behavior of two-layer neural networks. The first set of coupled difference equations corresponds to rotation-based forming and breaking of clustered hyperplanes:

$$\begin{cases}\Delta\varphi = A\sin(\varphi)\left[\beta_1^h - \beta_{crit} - \dfrac{\beta_{max} - \beta_1^h}{\beta_1^h} + \dfrac{\beta_{max} - \beta_{crit}}{\beta_{crit}}\right], \\ \Delta\beta_1^h = \beta_{max} - \beta_1^h\end{cases} \tag{11.3}$$

and the second corresponds to the translation-based mechanism

$$\begin{cases}\Delta dist = A\, dist\left[\beta_1^h - \beta_{crit} - \dfrac{\beta_{max} - \beta_1^h}{\beta_1^h} + \dfrac{\beta_{max} - \beta_{crit}}{\beta_{crit}}\right]. \\ \Delta\beta_1^h = \beta_{max} - \beta_1^h\end{cases} \tag{11.4}$$

In both sets of coupled difference equations:

β_{crit}	is to the value of β_1^h for which the clustered hyperplanes start to decluster
β_{max}	is the maximum attainable β_1^h for clustered hyperplanes.

In this section, mismatch between the actual derivative of the activation function (with respect to the weighted input) $f'(x)$ and the used "derivative" $\tilde{f}'(x)$ is analyzed using (11.3) and (11.4).

It is shown in sections 8.4 and 8.6 that the adaptation of the $\underline{\Delta W}^{\varepsilon}$ and of the $\Delta\beta_1^{bias}$, for any training example, is proportional to the difference between the $\tilde{f}'(x)$ of the clustered first-layer neurons. Hence, for translation-based clustering-declustering:

$$\Delta|\underline{\Delta W}^{\varepsilon}| \propto \tilde{f}'(\underline{W}_{1,i}\cdot\underline{U}_1)-\tilde{f}'(\underline{W}_{1,j}\cdot\underline{U}_1) \tag{11.5}$$

and for the translation-based mechanism

$$\Delta\Delta\beta_1^{bias} \propto \tilde{f}'(\underline{W}_{1,i}\cdot\underline{U}_1)-\tilde{f}'(\underline{W}_{1,j}\cdot\underline{U}_1). \tag{11.6}$$

In sections 8.4 and 8.6, the right hand sides of (11.5) and (11.6) were replaced by an expression that includes the second derivative of the activation function with respect to the weighted input. Doing so for (11.5) and (11.6),

$$\Delta|\underline{\Delta W}^{\varepsilon}| \propto \tilde{f}''(\alpha_1^h\beta_1^h+\alpha_1^{bias}\beta_1^{bias})\,(\underline{U}_1^F\cdot\underline{\Delta W}^{\varepsilon})$$

and

$$\Delta\Delta\beta_1^{bias} \propto \tilde{f}''(\alpha_1^h\beta_1^h+\alpha_1^{bias}\beta_1^{bias})\,\Delta\beta_1^{bias}\alpha_1^{bias}$$

For the $\tilde{f}''(x)$ function, it can be derived that:

$$\tilde{f}''(x) \approx \frac{f'(Gx+G\varepsilon)-f'(Gx-G\varepsilon)}{2\varepsilon} = Gf''(Gx) \tag{11.7}$$

According to (9.24), the activation function of first-layer neurons is activated in the middle region for training examples that result in significant weight adaptation for that neurons. More accurately, (9.24) states that for first-layer neurons for which the weight vector is significantly adapted: $f'''(x)<0$. An illustration is given in figure 11.5.

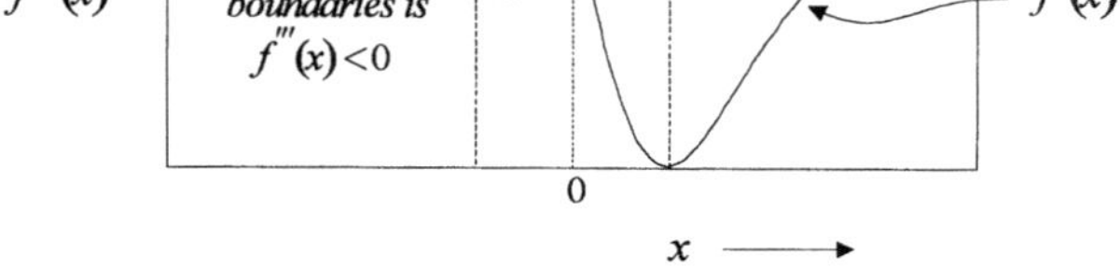

Fig. 11.5 For significant weight adaptation of first-layer neurons, the neuron is typically activated such that $f'''(x)<0$

It follows that the the first derivative of the first-layer neurons is a weak function of the weighted input for the region in which $f'''(x) < 0$, but that the second derivative is a much stronger function in this region. It can be concluded that *a mismatch as depicted in (11.2) mainly effects the first term between brackets in (8.31) and (8.47).*

Therefore, (11.2) and (11.3) can be modified as follows in order to obtain a sufficient approximation of the effect of mismatch as modelled in (11.2). The set (11.8) corresponds again to the rotation-based mechanism that leads towards the encounter and leaving of a temporary minimum:

$$\begin{cases} \Delta\varphi = A\sin(\varphi)\left[G'\left(\beta_1^h - \beta_{crit}\right) - \dfrac{\beta_{max} - \beta_1^h}{\beta_1^h} + \dfrac{\beta_{max} - \beta_{crit}}{\beta_{crit}}\right], \\ \Delta\beta_1^h = \beta_{max} - \beta_1^h \end{cases} \tag{11.8}$$

and (11.9) corresponds to the (related) translation-based mechanism

$$\begin{cases} \Delta dist = A\,dist\left[G'\left(\beta_1^h - \beta_{crit}\right) - \dfrac{\beta_{max} - \beta_1^h}{\beta_1^h} + \dfrac{\beta_{max} - \beta_{crit}}{\beta_{crit}}\right]. \\ \Delta\beta_1^h = \beta_{max} - \beta_1^h \end{cases} \tag{11.9}$$

Figure 11.6 shows three calculated phase trajectories using (11.8) for a certain set of initial conditions. The rightmost trajectory corresponds to a mismatch factor G smaller than one, while for the leftmost trajectory $G > 0$. Similar phase trajectories can be obtained for the translation-based

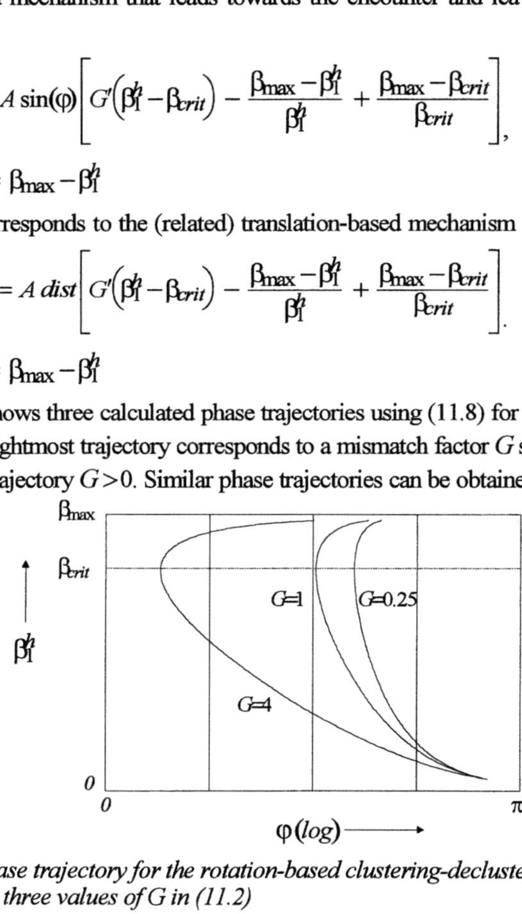

Fig. 11.6 Phase trajectory for the rotation-based clustering-declustering mechanism for three values of G in (11.2)

mechanism in (11.9).

It follows that the minimum angle φ decreases rapidly with G. This effectively means that the "sticking-time" of two-layer neural networks in temporary minima increases with G *and* that the (digital) resolution required to escape from the temporary minimum also increases with G.

This latter point can be explained as follows. For small minimum angles φ (or equivalently small minimum distances *dist*) the difference in response of clustered neurons is very small. However, this very small difference in response is used to cause a different weight adaptation $\Delta\underline{W}_{1,i}$ and $\Delta\underline{W}_{1,j}$ for the clustered hyperplanes in order to escape from the temporary minimum.

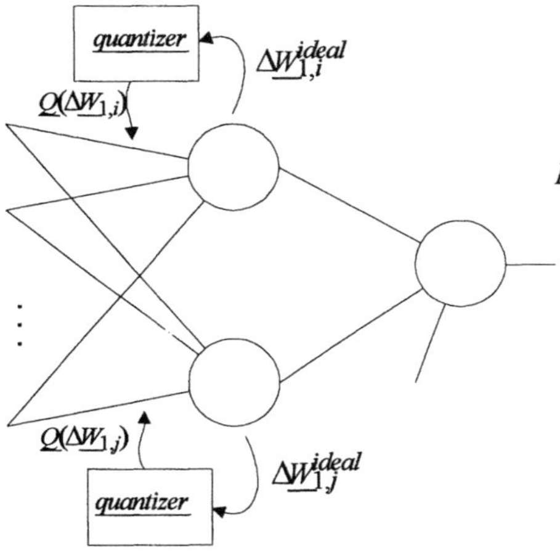

Fig. 11.7 If the weight adaptations of first-layer neurons are quantized, small "φ" and small "dist" cause infinitely long sticking in temporary minima

If the (digital) resolution of the weights and weight adaptation is now so small that every weight adaptation of the clustered neurons' weights are truncated to the same values, i.e. if for all examples $Q(\Delta\underline{W}_{1,i})=Q(\Delta\underline{W}_{1,j})$, the neural network is not able to escape from the temporary minimum. *In this case the temporary minimum is turned into a local minima due to lack of (digital) resolution.* Note that for limited analog resolution, this problem is not encountered.

Demands on the activation function for the second (output) layer

For the activation function in the last layer of a feed-forward neural network, basically the same demands apply as for neurons in single-layer networks (see section 11.2). Also for multi-layer neural networks, the derivative of the activation function of the last-layer's neurons can be omitted from the weight adaptation rule, which effectively results in using the cost-function (11.1).

Chapter 12

Feed-forward hardware

12.1 Introduction

Implementing neural networks in analog hardware seems natural because signal processing in neural nets is in principle in parallel with relatively simple operations. Additional advantages of neural network that are implemented in analog hardware is that:

- for sensory application one can directly connect the analog neural network to the (usually) analog sensors, which may even result in smart monolitic sensors
- high speed neural networks (either during learning or during operation after learning) can be realized
- analog hardware offers the possibility to implement small and low power (i.e. relatively slow) neural networks.

In section 12.2, the signals in the hardware neural network are normalized in order to properly adjust the learning rate η and the gain of synapses and activation functions. These topics are of main interest because the gain of the synapses and activations functions determines to a great extend the performance of the neural network (without learning) and because for on-chip learning neural nets the learning rate must be small enough to ensure stability during training while the η must be as large as can be tolerated in order to minimize parasitic effects (see for example chapter 6) and training time (see chapter 3).
The design of a feed-forward part of an on-chip learning neural network is presented in section 12.3 and 12.4. Extensive literature overviews with circuit schematics are provided in the appendices of this chapter. The next chapter presents the design (including literature overviews) of weight adaptation circuitry.

12.2 Normalization of signals in the network

Learning behavior of multi-layer feed-forward neural networks has been studied by several authors; in these papers, normalized signals are assumed. This means that among others the input and output of the neural network are in a unit range and that the maximum derivative equals one (or for a sigmoid $\frac{1}{4}$).

In hardware realizations of learning feed-forward multi-layer neural networks, any signal is generally not normalized. To be able to apply the existing theories and analyses to hardware implementations, it is therefore necessary to normalize the signals in the hardware neural network. In this section, the signals in the feed-forward neural network are normalized; the unit used are typical and will be used for the circuit shown in this chapter.

Feed-forward normalization

In figure 12.1, the feed-forward part of a two-layer neural network with its specific sub-blocks is drawn. For clarity of illustration, only two neurons are shown in the first layer and one neuron is shown the second layer. Proper normalisation of the signals in the hardware neural network is important because of two reasons. Firstly, when implementing off-chip learning, the hardwired weight range sould be set properly and the down-loaded weights must correspond to the intended weights which have been calculated off-chip. Secondly, for on-chip learning normalisation of all signals is important in order to properly adjust the learning rate η.

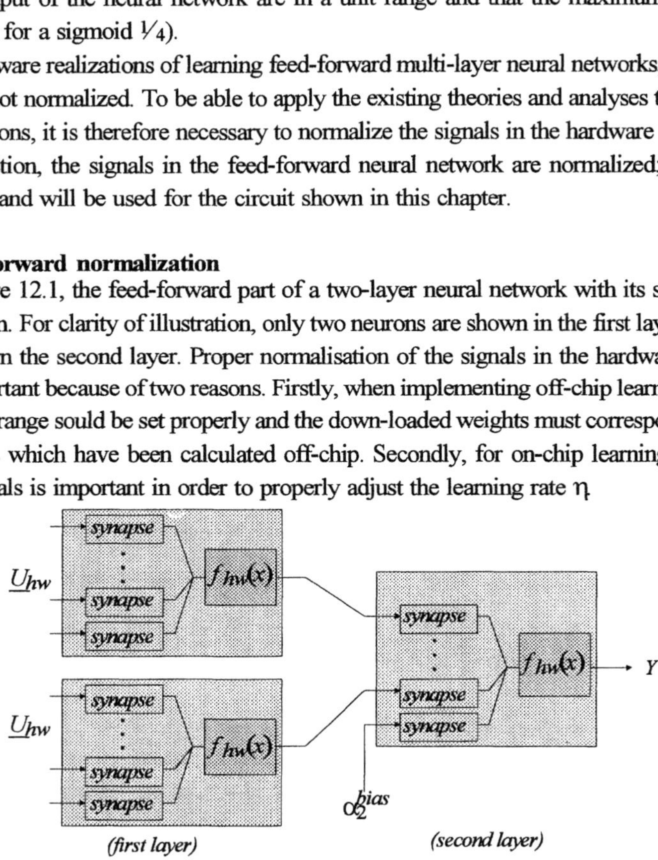

Fig. 12.1 Two-layer hardware neural network at block-level: feed-forward part

The feed-forward pass of a multi-layer neural network determines the response of the neural network on input vectors. The response of each neuron is a non-linear function on the weighted input of the neuron. Assuming sigmoid-shaped activation functions, this is

$$f_{hw}(x_{hw}) = f_{norm}(x_{norm}) \cdot S_{trm} = \frac{S_{trm}}{1+exp(-S_{trwi}\ x_{hw})} \tag{12.1}$$

where x_{hw} is the weighted input of the hardware activation function [A]

$f_{hw}(x_{hw})$ is the activation function of the hardware neural net

x_{norm} is the weighted input of the activation function for a network with normalized signals

$f_{norm}(x_{norm})$ is the activation function of a network with normalized signals

S_{trwi} is the weighted-input scale factor of the activation function $[A^{-1}]$

S_{trm} is the activation function output scale factor [A]

The width of the interval of the input signals will generally not be equal to one in hardware realizations of neural nets. Introducing a scale factor S_u for the input signals, the input vector for the hardware neural network is

$$\underline{U}_{hw} = \underline{U}_{norm}\, S_u \tag{12.2}$$

where $\underline{U}_{hw}$ is the input vector of the hardware neural network $[\underline{A}]$

$\underline{U}_{norm}$ is the normalized input vector of a neural network

S_u is the scale factor for the input signals [A]

In neural networks with normalized signals, synapses are implemented as scalar-multipliers with a gain factor equal to one. In hardware neural networks, synapses can be implemented as analog current-voltage multipliers with the output signal in the current domain; the gain of this multiplier is $S_{synapse}$:

$$S_{synapse} = \frac{I_{out}}{I_{in}\ V_{weight}}\ [V^{-1}] \tag{12.3}$$

Weights are usually represented by voltages across capacitors. The weight voltage corresponding to a unit weight will be denoted as S_w, and is given by:

$$S_w = \frac{1}{S_u\ S_{synapse}\ S_{trwi}}\ [V]. \tag{12.4}$$

In implementations of neural networks, it is efficient to have a unique neuron circuit block throughout the network. For analog hardware implementations this means that for example one synapse-circuit block and one sigmoid-activation block can be applied. Unless stated otherwise, it will be assumed that the blocks presented in this chapter can be applied in any layer.

Feed-back normalization

In this part of section 12.2, the signals in the feed-back part (or weight adaptation part) of feed-forward neural networks are normalized. Especially the normalization of this feed-back pass is important in on-chip learning neural nets because *too large* learning rates in the feed-back pass have been shown to result in not-converging. On the other hand, *relatively low* learning rates are harder to realize in analog or digital hardware and are increasing the required training time to reach a specific level of performance (see chapters 6 and 7 and [1]).

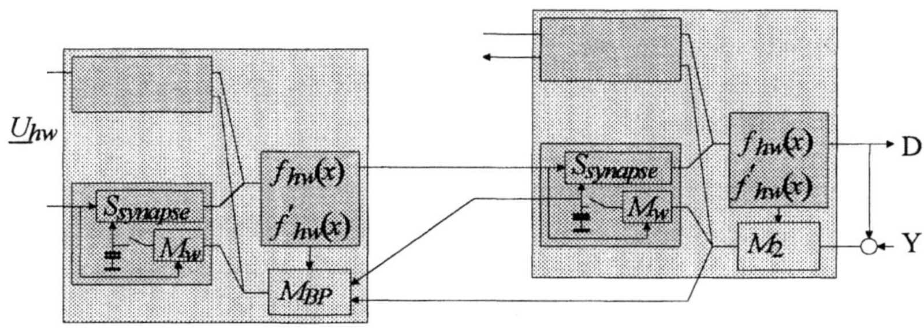

Fig. 12.2 Two-layer hardware neural network at block-level; one neuron per layer shown

The construction of a hardwired two-layer feed-forward neural network with back-propagation learning is presented in figure 12.2. In this realization, it is assumed that the weights are represented by voltages across integrated capacitors; weight adaptation can be either analog or digital using AD and DA converters.

For the neuron in the *second* layer, the block denoted as M_2 multiplies the error $(D-Y)$ and the first derivative of the activation function, with gain equal to M_2. The output of the M_w block in the second-layer neuron is the product of the input signal of the corresponding synapse, the output of M_2, and gain M_w:

$$(Y_{mw2})_i = M_w \cdot (\underline{U}_2)_i \cdot \left(M_2 \cdot f'_{hw}(\underline{U}_2 \cdot \underline{W}_2) \cdot (D-Y) \right) \tag{12.5}$$

The block denoted with M_{BP} multiplies the derivative of the corresponding neuron, the output of block M_2, the weight voltage, and a gain factor M_{BP}. The output of the M_w block of the corresponding neuron in the *first* layer is

$$(Y_{mw1})_i = M_w \; (\underline{U}_1)_i \left(M_{BP} \cdot f'_{hw}(\underline{U}_1 \cdot \underline{W}_{1,j}) \cdot (\underline{V}_{w2})_j \cdot M_2 \cdot f'_{hw}(\underline{U}_2 \cdot \underline{W}_2) \, (D-Y) \right). \tag{12.6}$$

For the weight adaptation of the weights, the derivative of the activation function is required.

With (12.1), this derivative can be written as

$$f'_{hw}(x_{hw}) = f'_{norm}(x_{norm})\, S_{dtrm} = \frac{S_{dtrm}}{1+exp(-S_{dtrwi}\, x)}\left\{1-\frac{1}{1+exp(-S_{dtrwi}\, x)}\right\}$$

where:

$f'_{hw}(x)$	is the derivative approximation
x_{hw}	is the weighted input of the hardware activation function [A]
x_{norm}	is the weighted input of the activation function for a network with normalized signals
$f'_{norm}(x)$	is the derivative of the neural network with normalized signals
S_{dtrwi}	is the weighted-input-scale factor of the derivative $[A^{-1}]$
S_{dtrm}	is the derivative-output scale factor [A]

Feed-back normalization: the last layer
In this section, the learning rate of the last layer of a hardwired feed-forward neural network is calculated. Note that for neural networks adapting according to the back-propagation algorithm (or a derivative) the weight adaptation of neurons in the last layer of two-layer neural network is equivalent to the weight adaptation of neurons in single-layer nets.
In a fully analog hardware neural network, the block M_W is implemented as an analog multiplier with current output. During adaptation, a switch gates this output current to the capacitor during an adaptation interval τ, resulting in a weight-voltage adaptation

$$(\Delta \underline{V}_{w2})_i = \frac{\tau}{C_{2,i}^{weight}}\,(D-Y)\,(\underline{U}_2)_i\, M_W M_2\, f'_{hw}((\underline{U}_x)_i \cdot \underline{W}_2)$$

where:

$(V_{w2})_i$	is the weight voltage corresponding to weight $(W_2)_i$ [V]
τ	is the adaptation time [s]
$C_{2,i}^{weight}$	is the value of the weight storage capacitor of the second layer neuron [F]
$D-Y$	is the error between desired and actual response of the neuron [A]
$(\underline{U}_2)_i$	is the input signal corresponding to weight $(\underline{W}_2)_i$ [A]

With the scale and gain factors previously introduced, it can be derived that the learning rate of the neuron in the last layer equals

$$\eta = \frac{\tau\, S_{trm}\, M_W\, S_u\, M_2\, S_{dtrm}}{C_x^{weight}\, S_W} \qquad (12.7a)$$

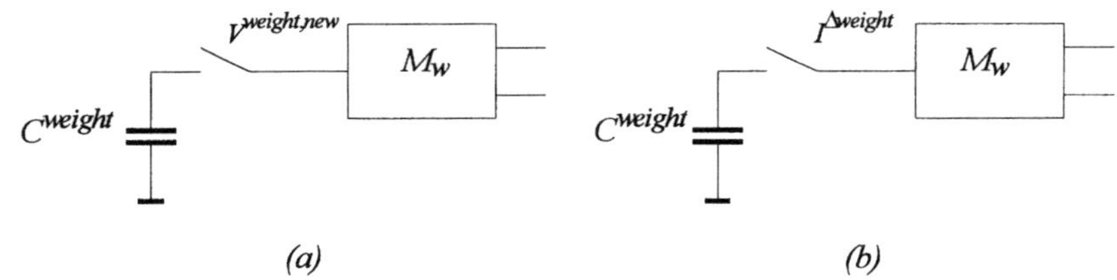

Fig. 12.3 Two ways to adapt weights: (a) load a new weight (typically using ADC and DAC) (b) fully analog weight adaptation through charge adaptation

When applying a digital based feed-back phase, M_W is implemented by a DA-converter whose output is usually the new weight voltage. During adaptation, the switch is closed in order to sample the voltage at the output of the DA-converter at the weight storage capacitor, see figure 12.3. The resulting learning rate is then

$$\eta = \frac{S_{trm}\ M_W\ S_u\ M_2\ S_{dtrm}}{S_W} \tag{12.7b}$$

The learning rate in (12.7b) remains dimension-less because the dimension of M_W is changed. It may be evident that when using a DA-converter with current output, (12.7a) is to be used.

Feed-back normalization: the first layer of a two-layer net

In two-layer feed-forward neural networks, the adaptation of the weights in the first layer is different from weight adaptation in the second layer, see (12.5), (12.6) and figure 12.2. For fully analog implementations of two-layer neural nets, it can be derived that the learning rate of the neurons in the first layer of the neural network is given by

$$\eta = \frac{\tau\ S_{trm}\ M_W\ S_u\ M_{BP}\ S_{dtrm}\ S_W\ M_2\ S_{dtrm}}{S_W\ C_{1,i}^{weight}} \tag{12.8}$$

For the forenamed digital-based feed-back phase using a DA-converter with voltage output, the adaptation time τ and the value of the weight storage capacitor $C_{1,i}^{weight}$ have to be omitted from (12.8); the learning rate again remains dimension-less because of the modified dimension of M_W.

12.3 Feed-forward hardware: the synapses

The response of a neuron in a feed-forward neural network is usually a non-linear function of the weighted input of the neuron. In formula,

$$Y = f(\underline{W} \cdot \underline{U}) = f\left(\sum_i \underline{W_i}\,\underline{U_i}\right) \tag{12.9}$$

The multiplication of input signals and weights is done by so-called synapses while the signal summing as well as the non-linear function are provided by the so-called activation function. A summary of demands on circuit implementations of synapses is presented in section 12.3.1. The design of a new type of multiplier is then presented in section 12.3.2. Extensive literature overviews are presented in Appendix 12.1.

12.3.1 Requirements

In feed-forward neural networks, a large number of synapses is required. This section presents first demands on synapses to be used in moderate speed analog hardware neural nets. Secondly, a synapse-circuit that satisfies the demands is presented.

Two-quadrant multiplier versus four quadrant multipliers

It has been shown in chapter 4 and 9 that for feed-forward neural networks, synapses can be implemented by two-quadrant multipliers. As two-quadrant multipliers consume less power and require less chip area than equally fast four-quadrant multipliers, we will use two-quadrant multipliers. Note that it may be necessary to include a level shift for the input signals to convert bipolar input signals to unipolar input signals. It was shown in chapters 4 and 9 that by proper adjustment of the bias input signal of neurons, the effect of shifting input signals in input space in order to obtain unipolar input signals can be minimized.

Single output versus symmetrical output

The number of input signals applied to the neuron, i.e. the dimension of the input vector, is generally large for neurons in a feed-forward neural networks. To be able to easily construct the weighted input of the neuron, $\underline{W} \cdot \underline{U}$, the output of the synapses are usually in the current domain, because signal summing in the current domain is simpy wiring all outputs together.

Basically, two different ways to derive the weighted input in hardwired neural nets are widely used. Firstly, all synapses have two symmetric output currents, which are first separately summed. The difference of the two sums is then the weighted input, figure 12.4a illustrates this method [3], [5].

Secondly, all synapses have one bipolar output current and the weighted input is then the sum of these bipolar output currents, see figure 12.4b. This second method generally requires a larger chip area for each synapse because each synapse usually includes a current mirror. The latter

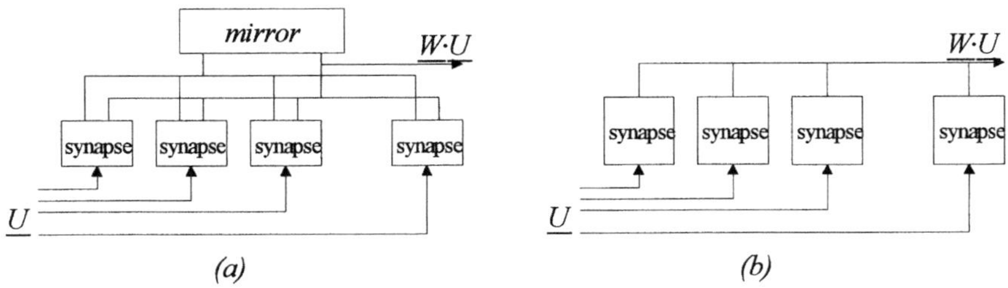

Fig. 12.4 Two weighted input configurations: (a) inhibitory and excitatory current summing and extraction and (b) single output current summing

approach has as advantages that the wiring is more simple compared to the method in figure 12.4a and that all building blocks are identical, i.e. the current mirror(s) do not need to be redesigned when implementing a neuron with a higher or lower dimensional input vector. Assuming that the current mirrors in figures 12.4a and 12.4b have identical current transfer (generally not unity because of the Early-effect), the *accuracy of both methods is identical.*

However, some the total required chip are for both methods is different as will be shown below. In feed-forward neural networks, the weighted input $\underline{W{\cdot}U}$ is fed to a non-linear function block. Assuming that this non-linear function is approximately saturated for weighted inputs satisfying $|\underline{W{\cdot}U}| > \underline{WU}_{sat}$, the weighted input need only to be relatively accurate for a limited range:

$$\begin{cases} \underline{W{\cdot}U} < -\underline{W{\cdot}U}_{sat} & \textit{non-accurate summing} \\ |\underline{W{\cdot}U}| < \underline{W{\cdot}U}_{sat} & \textit{limited-accurate summing} \\ \underline{W{\cdot}U} > \underline{W{\cdot}U}_{sat} & \textit{non-accurate summing} \end{cases} \tag{12.10}$$

Note that in the "limited accuray" range the summing can be limited non-linear because this non-linearity can simply be accredited to the non-linear activation function. Relation (12.10) states that for both methods in figure 12.4 any mirror need to be relatively accurate for $|\underline{W{\cdot}U}| < \underline{WU}_{sat}$.

Denoting the bias current components in the two symmetric output currents of the synapse (before the subtraction) as I_b, the single current mirror in figure 12.4a must then be designed to operate accurately on at least $N_{in}{\cdot}I_b + \frac{1}{2}\,\underline{WU}_{sat}$ whereas all mirrors in figure 12.4b must be dimensioned to accurately operate on at least $I_b + \frac{1}{2}\underline{WU}_{sat}$. As it is not necessary to dimension the synapse in such a way that the output current of the synapse significantly exceeds $\underline{WU}_{sat}$, the condition $I_b \approx \frac{1}{2}\,\underline{WU}_{sat}$ is generally satisfied. It follows that the ratio between area consumption of the single mirror in figure 12.4a and of the N_{in} mirrors in figure 12.4b is

$$\frac{A_N}{A_1} = \frac{N_{in}{\cdot}I_b + \frac{1}{2}\,\underline{WU}_{sat}}{N_{in}{\cdot}(I_b + \frac{1}{2}\underline{WU}_{sat})}. \tag{12.11}$$

For very large N_{in}, this factor equals 2.

It can hence be concluded that one single mirror is much more efficient than N_{in} mirrors that individually mirror the output currents of synapses. By distributing the single mirror in figure 12.4a over all synapses, these distributed mirrors can be smaller than the corresponding mirrors in figure 12.4b by a factor up to 2. Another advantage of distrubuting the mirror in 12.4a over all synapses is that no redesign of the mirror is required when implementing a neuron with input vector having another dimension.

Required transconductance and input impedance

In on-chip learning neural networks, the weights have to be adapted. In fully analog implementations of the forward-part of the neural network, the weights are usually represented by voltages stored on integrated capacitors. To relax the accuracy demands for the weight-adaptation-circuits, a large voltage swing across the capacitors is required (see chapter 6). This effectively means that the transconductance from weight-voltage to output current of the synapse must be low. Also when applying a digital (based) weight-adaptation pass, or in case of downloading precalculated weights as analog voltages, it is advantageous to have a small transconductance in order to minimize the sensitivity for weight-errors. Other demands on the synapse-block resulting from using capacitive weight storage include a weight-signal input which draws no DC-current.

Type of multiplier to be used

In analog hardware realizations of neural nets, synapses are usually implemented as analog multipliers. Because a large number of synapses is required in neural networks, it is important to minimize the chip area and power consumption required to implement one single synapse. However, one must keep in mind that the network functionality must be preserved in the strive to minimize area and power consuption.

In literature, small analog multipliers for operation in feed-forward neural networks are realized in many different ways. An overview (including circuit schematics) is provided in Appendix 12.1 and a short discussion of a number of multipliers is given below:

- The Gilbert Cell based multipliers perform a four-quadrant multiplication. Because only two-quadrant multiplications are required they will usually not lead to minimum sized and powered circuits.

- Analog multipliers using the difference of squares principle generally have a high linearity and are small. However, two low-ohmic nodes are required to sum the squares and one current mirror must be used to subtract the sum of squares. In hardware, low ohmic nodes can only be realized for relatively low frequency applications.

- The MOS-triode based multipliers are similar to the multipliers that realize the difference of squares.

- The translinear multiplier operates in the current domain and is hence difficult to implement in neural nets in which weights are stored as voltages on capacitors.
- For differential pairs, the differential voltage must be used for the weight and the biasing then corresponds to the input signal. The major problem using differential pairs in neural nets is the low linearity which may degrade learning significantly.

In the following section, a differential-pair based analog multiplier is presented that can be used in moderate speed learning neural nets. This multiplier has sufficient linearity and does not require low-ohmic current summing nodes nor draws current from the weight-voltage inputs.

12.3.2 The synapse circuit

The multiplier to be used in moderate speed *on-chip learning* multi-layer feed-forward neural networks must satisfy a number of demands. The most important are listed below.

- the multiplier must be fast enough to operate on moderate speed (low MHz. range),
- the circuit must have a symmetrical output current,
- the input voltage swing (to be used for the weight) must be large (see chapter 6)
- the linearity of the multiplier must be relatively good in order to not-degrade learning, and
- the *bipolar* input signal of the multiplier (to be used for the weight) must draw no (DC type) current because weights are usually stored as voltages across capacitors.

Furthermore, it was shown in chapter 4 and 9 that two-quadrant multiplications are sufficient in feed-forward neural nets. Because of the demand for a low transcunductance (the third demand), it appears that both multipliers based on MOS triode operation and differential pairs with significant source degeneration can be used as synapses in *on-chip learning* feed-forward neural nets. An overview of these types of multipliers can be found in the appendix. In this section, we will introduce a new MOS triode multiplier, which can also be used in analog on-chip learning feed-forward neural nets.

Note that for *off-chip learning* feed-forward neural networks, it may appear optimum to use either analog multipliers with EEPROM (analog non-volatile) memories to store weights, or to use multipliers with inherent DA-conversion, see appendix 12.1.

A new MOS triode multiplier

The proposed new multiplier is a mixture of the MOS differential pair and the MOS triode circuit of Reed and Geiger (see appendix 12.1). The circuit is presented in figure 12.5. In this multiplier, the two MOS transistors that are connected to the weight voltage V^{weight} are operated in triode region. The drain-source voltage is controlled by the two cascode transistors M_{c1} and M_{c2} and the MOS diode M_d.

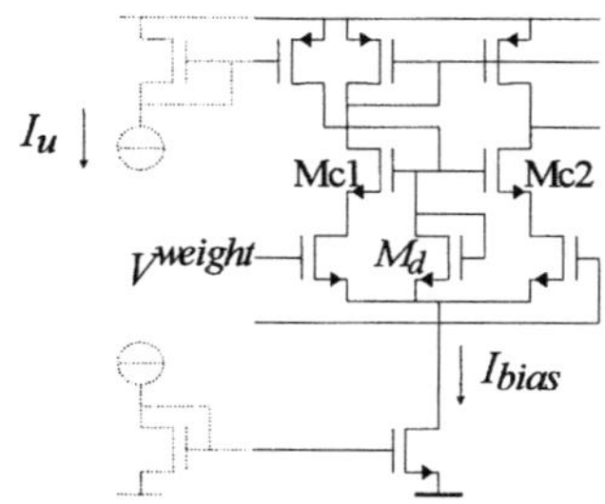

I_{ref} (connect to all other synapses' I_{ref})

I_{out} (connect to all other synapses' I_{out})

Fig. 12.5 Synapse: linearized differential pair based on MOS triode operation

The two input transistors operate in triode region, which results in a relatively low transconductance from input voltage V^{weight} to output current. Furthermore, the drain-source voltages of the MOS transistors in the differential pair are modulated by the voltage across the MOS-diode M_d via the two cascode transistors M_{c1} and M_{c2}. The current through the MOS-diode M_d is the (unipolar) input current I_u which current requires a non-zero bias component to forward-bias the MOS-diode. Figure 12.5 shows the synapse circuit with a distributed mirror as discussed in section 12.3.1.

Note that the voltages at the two voltage-input nodes of the multiplier are quasi-static during training and are fixed if the neural network is not adapting. Because of this quasi-static nature of V^{weight}, the input voltage takes care of a quasi-DC multiplication factor of the input current I_u.

The weight-range achievable by this synapse depends on the slope of the activation function used, see (12.4). Using the activation circuit that is presented in 12.4, it follows from (12.4) that for the MOS-transistor dimensions and process chosen, $S_w \approx 9.6V^{-1}$ (related to a neural network with normalized signals, sigmoids with a maximum derivative equal to 1.0, and unity gain for any multiplier). This means that a differential weight voltage equal to $^{1.0}/_{9.6}$ V corresponds to a unit weight in the specified neural network with normalized signals.

Simulation results
The simulated output current of the synapse of figure 12.5 is plotted in figure 12.6 as a function of the input current for weight voltages in the range $[-1.0V, 1.0V]$.

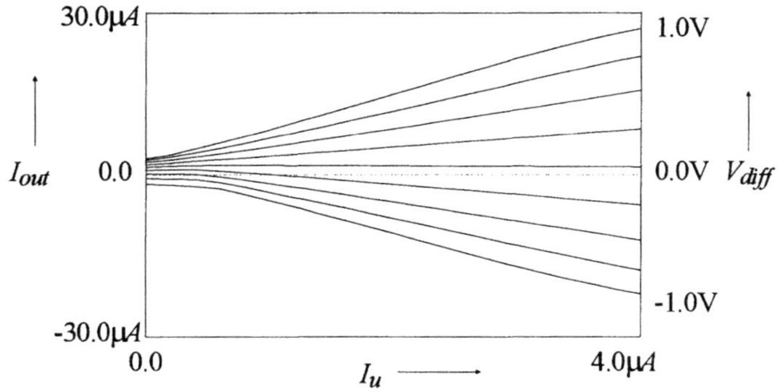

Fig. 12.6 DC-transfer of the synapse for weight voltage ranging from -1.0 to 1.0V

Within the input signal range $I_u \in [0.0\mu A, 4.0\mu A]$, the linearity is better than 9.3% of the maximum output current (average 1.3%), which is sufficient for application of this multiplier in on-chip learning feed-forward neural networks. Furthermore, PSpice simulations show that the rise and fall time is approximately $60ns$ which is sufficient for moderate speed operations.

12.4 Feed-forward hardware: the activation function

In feed-forward neural networks, the response of neurons is usually a non-linear function of the weighted input. The weighted input is realized by summing the bipolar output currents of the synapses on a summing node; the non-linear function is to be realized by the activation function circuit. Furthermore, for fully analog on-chip learning neural networks using back-propagation, the derivative of the non-linear function with respect to the weighted input is required. Requirements on the activation function and its derivative are discussed in chapter 11.

In hardware neural networks, the non-linear function is usually realized by either a CMOS differential pair or by inverter-based circuits; an overview is presented in appendix 12.2. The differential pair-based circuits typically have a current output signal while the output of the inverter-based circuits is in the voltage domain. On-chip learning feed-forward neural networks require circuitry for the derivative of the non-linear function; most existing designs described in literature (see appendix 12.2) are based on the fact that for the sigmoid function:

$$f'(x) = f(x)\left(1 - f(x)\right). \tag{12.12}$$

These implementations use the output signal of the non-linear function and a multiplier to construct an approximation of the actual derivative. Another analog hardware implementation of the derivative which calculates a type of numerical derivative, has recently been published and is shown in appendix 12.2. In this section, a novel CMOS differential pair based non-linear function, including a relatively simple derivative approximating circuit, is presented.

Design of the non-linear function circuit

Probably the most widely used non-linear function used in feed-forward neural networks is the so-called sigmoid function. The sigmoid function can be approximated very well by using a bipolar differential pair; the input variable being the differential input voltage and the output variable being one of either output currents. In CMOS processes the output current of a CMOS differential pair operated in weak inversion approximates the sigmoid as well [4]. For this circuit, the output current is

$$I_d = I_{bias}\left(\frac{1}{2} \pm \frac{1}{2}\tanh\left\{\frac{e\,V_{in}}{2\,n\,k\,T}\right\}\right) \tag{12.13}$$

where:

I_{bias}: the bias current of the differential pair [A]

V_{in}: the differential input voltage [V]

n: $1 + C_{dep}/C_{ox} \approx 1.5$

The circuit and the corresponding simulations are presented later on in this section.

Design of the derivative circuit

The existing derivative approximating circuits either multiply the drain currents of both MOS transistors in the differential pair or perform a type of numerical derivation to approximate the derivative of the sigmoid. In these ways, the derivative approximator is a relatively large circuit, while the approximation is not very accurate.

Another way to construct the derivative of the sigmoid function is to use the fact that outside the middle region $f(x)$ is either almost maximum or almost minimum. In these regions, the derivative $f(x)\left(1-f(x)\right)$ can then be approximated by the minimum of $f(x)$ and $1-f(x)$. An illustration of this approximation is given in figure 12.7.

It may be clear from the figure that the minimum of the two outputs of a MOS differential pair in weak inversion, $f(x)$ and $(1-f(x))$, is too large in the middle region of the sigmoid to satisfactory approximate its derivative. This can be solved by decreasing the gain of the minimum selector in the middle region of the sigmoid.

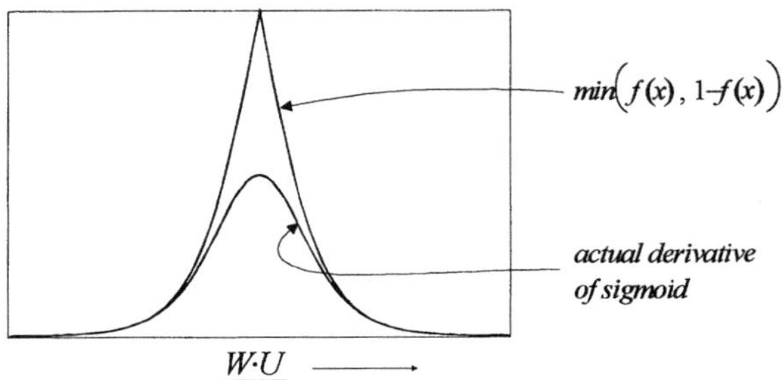

Fig. 12.7 The actual derivative and its "$min(f(x), 1-f(x))$" approximation

Figure 12.8 shows a hardware design of an activation function based on MOS-weak inversion operation with minimum-selector circuitry to approximate the derivative. The output current of the non-linear function block is denoted as I_f and the current that approximates the derivative is called I_{fder}. The non-ideal minimum selectors are basically two current mirrors with their output MOS-transistors in a type of cascode configuration. In this way, if the difference between the two input currents is relatively large, the output current of one minimum selector circuit is approximately the minimum of the two input currents. If the input currents are close, the M_{3a} and M_4 force respectively M_3 and M_{4a} into the linear region which decreases the output current in the middle region of the sigmoid. Two non-ideal (non-symmetrical) minimum selectors in a symmetric configuration are used to obtain a symmetrical approximation of the derivative.

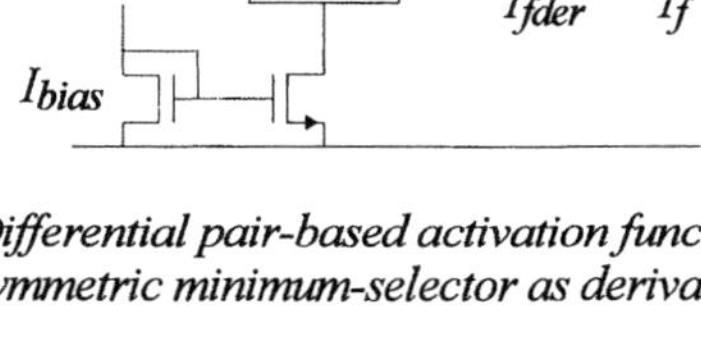

Fig. 12.8 Differential pair-based activation function and symmetric minimum-selector as derivative (dotted)

Design of current summing IV converter

The output signal of the synapses is in the current domain. Because the input signal of the non-linear function and of the derivative approximating circuit is in the voltage domain, an IV converter must be used to convert the input current into voltage. Furthermore, because the non-linear function circuit in figure 12.8 has a differential input signal, another IV converter with zero input current must be used to properly bias the non-linear function block. Demands on the IV-converters include low ohmic input resistance with respect to the output resistance of the synapses, the input node at a non-zero voltage for a proper operation of the synapses and a non-zero bias output voltage for biasing of the circuit in figure 12.8.

All demands can be satisfied by using an IV-converter based on Gatti-VI-converters [2]. These Gatti-VI-converters are based on MOS-triode operation of one transistor with a constant drain-source voltage. The constructed IV-converter is shown in figure 12.9. (Another way is to use the current mirror configuration in figure A12.3 in appendix 12.1).

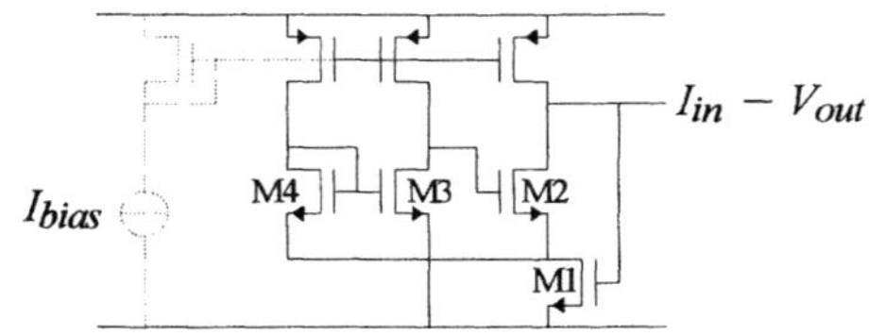

Fig. 12.9 Gatti-VI converter used as IV converter.

For this IV-converter, it can be derived that the transresistance depends on two bias currents and on the sizes of three NMOS transistors. Of these transistors, M3 and M4 operate in saturation and adjust the drain-source voltage of M1 (via M2). In equation,

$$\frac{\partial I_{in}}{\partial V_{in}} = \frac{K}{A}\frac{W_1}{L_2}V_{ds1} = K\frac{W_1}{L_1}\,\frac{\sqrt{2\frac{I_4}{K}\frac{L_4}{W_4}} - \sqrt{2\frac{I_3}{K}\frac{L_3}{W_3}}}{A} \tag{12.14}$$

In this equation, the A parameter corresponds to the body effect parameter [42].

The activation function circuit

The activation function circuit consists of two IV converters, the non-linear function circuit and of the derivative approximation circuit. One of the two IV-converters is used to sum and convert the input current, the other IV-converter is used to only bias the differential pair:

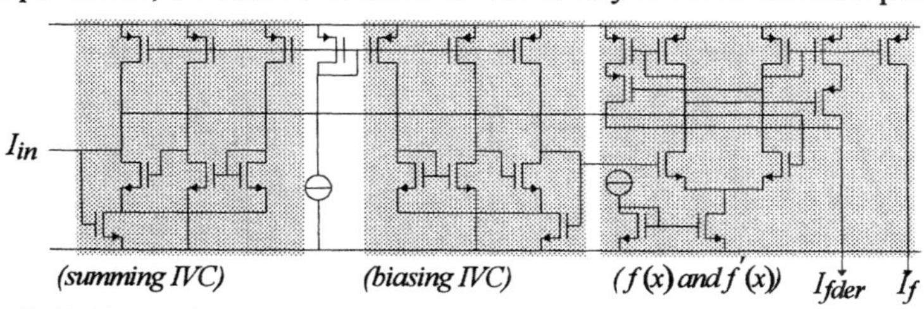

Fig. 12.10 The total activation function circuitry

A number of PSpice simulations and curve-fits on the results from these simulations are done in order to verify the correct operation of the circuit in figure 12.10. The output of the transfer function circuit should approximate a scaled version of the sigmoid function. The simulated transfer is therefore fitted on a sigmoid function, resulting in

$$I_f \approx \frac{S_{trm}}{1+\exp(-S_{trwi} * I_{in})} \tag{12.15}$$

with: $S_{trm} = 3.70\ \mu A$

$S_{trwi} = 2.08\ (\mu A)^{-1}$

The maximum error between the PSpice simulated output signal of the hardware transfer function and the fitted sigmoid curve is 2% of the maximum response of the activation function block. Two possible fits can be performed on the simulated output current of the derivative approximating circuit: the simulated output current can be fitted on the derivative of a true sigmoid and the simulated output current can be fitted on the derivative of the simulated I_f current. The differences between both methods are minor and in this section, the PSpice simulated response of the derivative approximating circuitry is fitted on the derivative of the true sigmoid. The result of this fit is

$$I_{fder} \approx \frac{S_{dtrm}}{1+exp(-S_{dtrwi}*I_{in})}\left\{1-\frac{1}{1+exp(-S_{dtrwi}*I_{in})}\right\} \tag{12.16}$$

with: $S_{dtrm} = 13.1\ \mu A$

$S_{dtrwi} = 2.19\ (\mu A)^{-1}$

The maximum error between the simulated output of the hardware realization of the derivative approximator and the fitted curve is 5% of the maximum of the derivative. Figure 12.11 shows the two simulated output signals of the transfer circuit, and shows the corresponding fitted curves.

Although the matching between actual and hardware-constructed derivative can be improved (especially for the matching of the S_{trwi} and the S_{dtrwi} parameters) this is not done because the required chip area would increase significantly while the present matching is satisfactory.

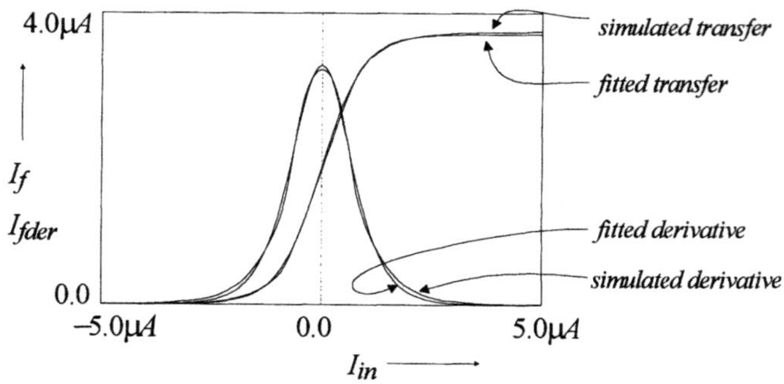

Fig. 12.11 Simulated and fitted curves of the activation function and its derivative

12.5 Conclusions

In this chapter, analog circuits to be used in the *feed-forward* part of *on-chip learning* feed-forward neural networks are presented. The demands on circuits for on-chip learning feed-forward neural networks are different from demands for non-learning feed-forward neural networks. This is caused by the fact that for on-chip learning networks, demands imposed by weight adaptation hardware (see chapter 6) must be included in the design of the feed-forward part of the network. For non-learning neural networks, generally different demands apply: the speed, power consumption and ease of programming the neural network are then of main interest.

Both for on-chip learning and for non-learning neural networks, normalization of signals has to be done in order to either implement proper learning or to implement a desired function. The first part of this chapter presents a normalisation of both the feed-forward and the feed-back(i.e. weight adaptation) part of neural networks.

Sections 12.3 and 12.4 present novel synapse and activation circuits which are suited for usage in on-chip learning multi-layer feed-forward neural networks. The appendices corresponding to these sections present literature overviews, including circuit schematics and short discussions, of existing analog hardware designs for synapses and activation functions.

For on-chip learning feed-forward neural nets, weight adaptation circuitry is required. The design of proper weight adaptation circuitry is the subject of the next chapter.

12.6 References

[1] H.A.C.Eaton, and T.L.Oliver, "Learning Coefficient Dependence on Training Set Size", Neural Networks, vol. 5, pp. 283-288, 1992

[2] U.Gatti, F.Maloberti and G.Torelli, “A Novel CMOS Linear Transconductance Cell for Continuous-Time Filters”, in Proc. ISCAS 1990, New Orleans, pp. 1173-1176

[3] J.J.Hopfield, “The effectiveness of analogue ’neural network’ hardware”, Network, vol. 1, pp. 27-40, 1990

[4] C.Mead, “Analog VLSI and Neural Systems”, New York: Addison-Wesley, 1989

[5] M.Verleysen, P.Jespers, “Precision of Sum-of-Products in Analog Neural Networks”, in Proc. 1st Int. MicroNeuro, Dortmund 1990, pp. 47-61

[6] H.Wallinga, “De MOS transistor”, University of Twente, kenmerk 1217.2225, 1978

Appendix 12.1: Neural multipliers: overview

In this appendix, an overview of some literature that describes mutlipliers that are used (or can be used) in feed-forward neural networks is presented. This overview is not exhaustive but meant to give relatively fast a good overview. To limit the length of this overview, strongly related implementations are grouped; the actual implementation may for example have different current mirrors or may apply EEPROM devices instead of MOS transistors.

Differential pairs

Differential pairs can be used to implement (non-linear) two-quadrant multipliers. Among others Pelayo presented and discussed three synapses which are based on differential pairs. In the paper some advantages and disadvantages are discussed of the three approaches.

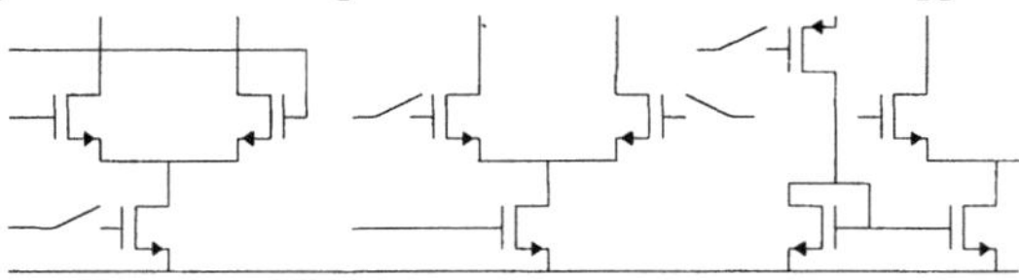

Fig. A12.1 Synapses based on signle differential pairs (switches used to load weights)

[1] *F.J.Pelayo, B.Pino, A.Prieto, J.Ortega, and F.J.Fernandez, "Analog Implementation of a Synaptic Matrix with Alterable Weights", in Proc. NeuroMicro 1991, pp. 261-268*

[2] *J.Ghosh, P.Lacour and S.Jackson, "OTA-Based Neural Network Architectures with On-Chip Tuning of Synapses", IEEE tr. CAS II, 1994, pp. 49-58*

[3] *R.L.Walke, S.F.Quigley and P.W.Webb, "Design of an anlogue subthreshold multiplier suitable for implementing an artificial neural network", IEE Proceedings - G, vol.139, no. 2, 1992, pp. 261-264*

[4] *J.B.Lont and W.Guggenbuhl, "Analog CMOS Implementation of a Multilayer Perceptron with Nonlinear Synapses", IEEE tr. Neural Networks, vol. 3, 1992, pp. 457-465*

Symmetrical differential pair configuration

A number of analog implemented feed-forward neural networks use differential pairs in a symmetrical configuration. In the circuit design by Satyanarayana, the lower the MOS transistors are ordinary MOS transistors. The weights are stored as (volatile) analog voltages on the gates of these MOS transistors. Downloading of the weights is done using MOS switches. In the widely known Intel ETANN chip, a variant of this synapse is used. The Intel ETANN chip was designed to operate with off-chip or with "chip-in-the-loop" method. To implement non-volatile weights, the lower two MOS transistors are implemented as EEPROM devices.

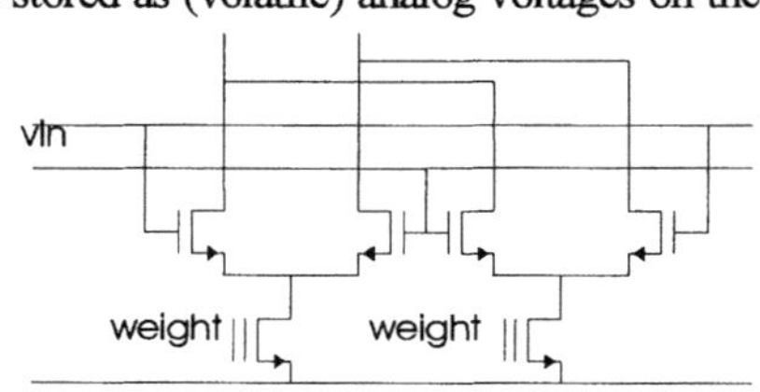

Fig. A12.2 Symmetrical differential pair synapse

[5] *Y.Tsividis and S.Satyanarayana, "Analogue Circuits for Variable-Synapse Electronic Neural Networks", Electronics. Letters, 1987, pp. 1313-1314*

[6] *F.J.Pelayo, B.Pino, A.Prieto, J.Ortega, and F.J.Fernandez, "Analog Implementation of a Synaptic Matrix with Alterable Weights", in Proc. NeuroMicro 1991, pp. 261-268*

[7] *H.A.Castro, S.M.Tam and M.A.Holler, "Implementation and Performance of an Analog Nonvolatile Neural Network", Analog Integrated Circuits and Signal Processing, 4, 1993*

[8] *M.A.Holler, "An Electrically Trainable Artificial Neural Network (ETANN) with 10240 Floating Gate Synapses", IJCNN, pp. 191-196, 1989, Washington*

Another symmetrical differential pair configuration

In 1990, Kub et al. presented a circuit which is very similar to the circuit of figure A12.2. The circuit also consist of two diferential pair in a symmetrical configuration. The main difference with the previous circuit is that the weight and inputs are interchanged.

With this circuit, the weight voltage range is relatively small, and this circuit is henceforth not very well suited for application in on-chip learning neural nets nor in neural network for which the weights are downloaded from an external host computer.

[9] *F.J.Kub, K.K.Moon, I.A.Mack and F.M.Long, "Programmable Analog Vector-Matrix Multipliers", IEEE JSSC, 1990, pp. 207-214*

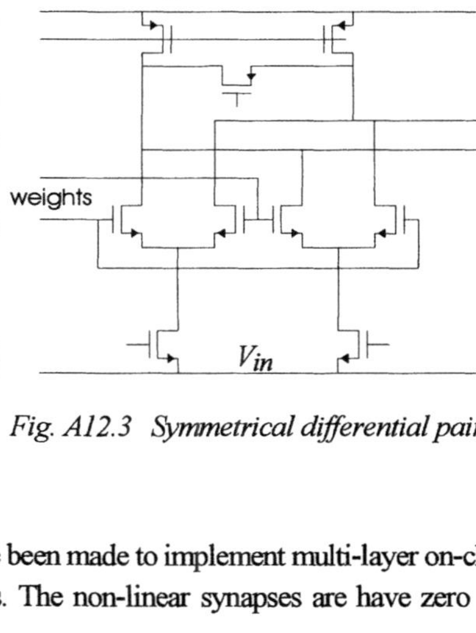

Fig. A12.3 Symmetrical differential pair

Intentionally non-linear synapses

Especially at the university of Genoa, attempts have been made to implement multi-layer on-chip learning neural networks with non-linear synapses. The non-linear synapses are have zero response for small weights and are relatively linear for large weights.

This non-linear synapse is reported to reduce power consumption, because for any small weight the power consumption is zero.

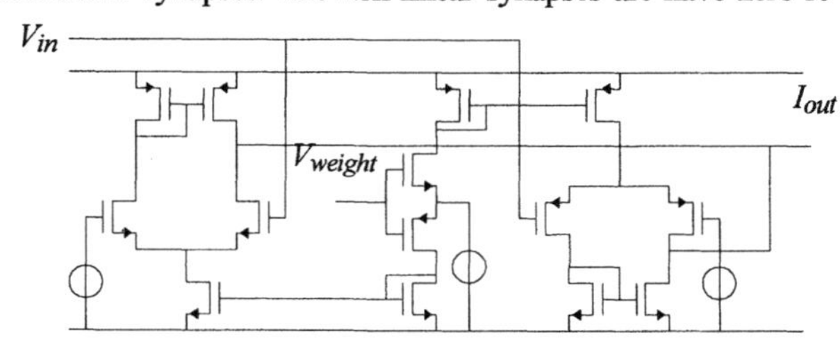

Fig. A12.4 Intentionally non-linear synapse

[10] *M.Valle, D.D.Caviglia, and G.M.Bisio, "An Experimental Analog VLSI Neural Chip with On-Chip Back-Propagation Learning", in Proc. ESSCIRC 1992, Copenhagen, pp. 203-206*

[11] *D.D. Caviglia, M.Valle and G.M.Bisio, "A VLSI Module for Analog Adaptive Neural Architectures", in VLSI 91, eds. Halaas and Denyer, 1992*

Differential pairs with digitally adjustable gain

For analog hardware feed-forward neural networks that are trained off-chip or "in-the-loop", typically a DA-converter is used to transfer the weights to the chip. For a number of realisations, this DA-conversion is done off-chip, which results in on-chip storage of analog weights. Some other realisations are resported to use on-chip DA-conversion, which is typically done using analog multipliers with a binary tail current as shown in figure A12.5.

Negative weights can be achieved by reversing the polarity of the differential input signal using four MOS switches. The "mirror" circuit consists basically of two current sources and a MOS transistor operating in its linear region. The current sources require common-mode feed-back in order to keep the differential output voltage at a certain common mode level.

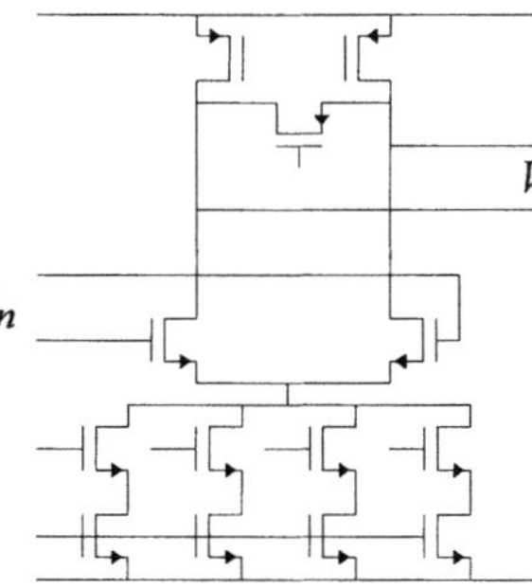

Fig. A12.5 Synapse with digitally adjustable weight (i.e. including a DAC)

[12] P.W.Hollis, "Artificial Neural Networks Using MOS Analog Multipliers", IEEE JSSC, 1990, pp. 849-855

[13] T.Duong, S.P.Eberhardt, M.Tran and A.P.Thakoor, "Learning and Optimization with Cascaded VLSI Neural Network Building-Blocks", in Proc. IJCNN 1992, Baltimore, vol. 1, pp. 184-189

[14] J. Van der Spiegel, P.Mueller, D.Blackman, P.Chance, C.Donham, R.Etienne-Cummings and P.Kinget, "An Analog Neural Computer with Modular Architecture for Real-Time Dynamoc Computations", IEEE J. Solid State Circuits, vol. 27, pp. 82-92, 1992

[15] P.H.W.Leong and M.A.Jabri, "Kakadu, A Low Power Analogue Neural Network Classifier", Int. Journal of Neural Systems, vol. 4, 1993, pp. 381-394

[16] P.Masa, K.Hoen and H.Wallinga, "70 Input, 20 Nanosecond Pattern Classifier", in Proc. IJCNN 1994, Orlanda Fla.

[17] P.W.Hollis and J.J.Paulos, "A Neural Network Learning Algorithm Tailored for VLSI Implementation", IEEE tr. Nueral Networks, vol. 5, pp. 784-791, 1994

Switches capacitor synapses

Some people made attempts to implement switched capacitor neural networks. In these SC neural networks, switched capacitors replace resistors. There are very few SC-realisations of neural networks, but they are shown here for sake of completeness.

[18] A.Mekkaoui and P.Jespers, "Four Quadrant Multiplier for Neural Networks", El. Letters, 1991, pp. 320-322

[19] Y.P.Tsividis and D.Anastassiou, "Switched-Capacitor Neural Net- works", Electronics Letters, 1987, pp. 958-959

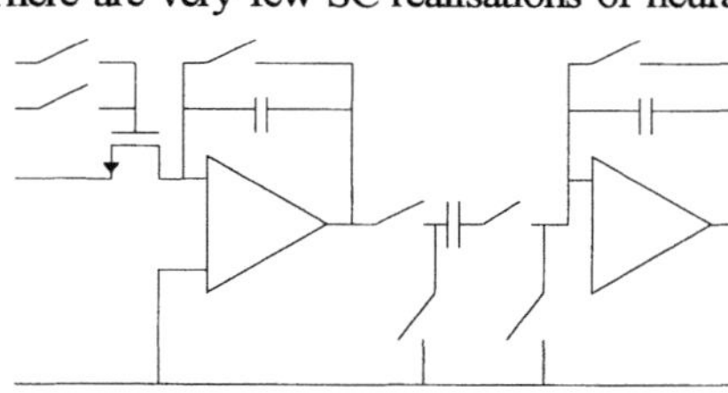

Fig. A12.6 Switched Capacitor synapse

Gilbert Cell

A well known four-quadrant multiplier is the Gilbert Cell, which was initially proposed for bipolar implementations. A widely known variant of this Gilbert Cell uses MOS transistors instead of the original bipolar ones. Among others Satyanarayana proposed to use this CMOS version of the Gilbert Cell as a synapse in neural networks.

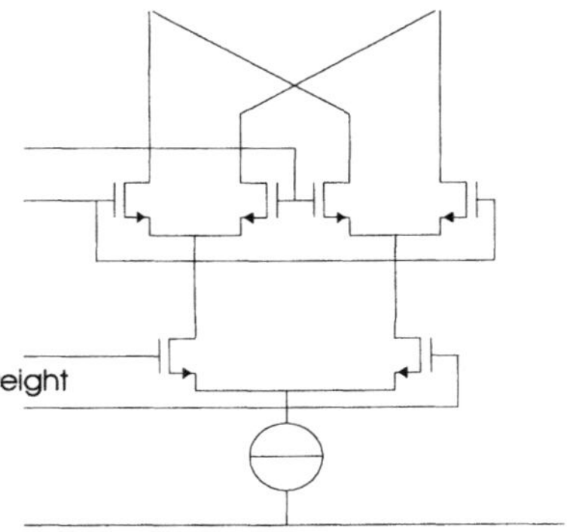

Fig. A12.7 Gilbert-cell based synapse

[20] *S.Satyanarayana, Y.P.Tsividis ans H.P.Graf, "A Reconfigurable VLSI Neural Network", IEEE JSSC, 1992, pp. 67-81*

[21] *J.Choi and B.J.Sheu, "VLSI Design of Compact and High-Precision Analog Neural Network Processors", in Proc. IJCNN, 1992, Baltimore, vol. 2, pp. 637-641*

[22] *B.Linares-Barranco, E.Sanchez-Sinencio, A.Rodriguez-Vazquez and J.L.Huertas, "Modular T-Mode Neural Network Learning Hardware Implementations with Analog Storage Capability", in Proc. IJCNN 1992, Baltimore, vol. 1, pp. 202-207*

[23] *Y.Wang, "A Modular Analog CMOS LSI for Feedforward Neural Networks with On-Chip BEP Learning", in Proc. IEEE ISCAS 1993, Chicago, pp. 2752-2755*

[24] *W.C.Fang, B.J.Sheu, O.T.C.Chen and J.Choi, "A VLSI Neural Processor for Image Data Compression Using Self-Organization Networks", IEEE tr. Neural Networks, vol. 3, 1992, pp. 506-517*

MOS triode synapses

A large number of relatively linear multuliers are based on MOS triode operation. In this region of operation, the drain-current of MOS transistors is by approximation proportional to the gate-source and drain-source voltage. In symmetrical configurations, MOS triode multipliers are inherently linear (when neglecting second order effects such as mobility reduction). The most widely known MOS triode multiplier was designed by Czarnul and Khoury.

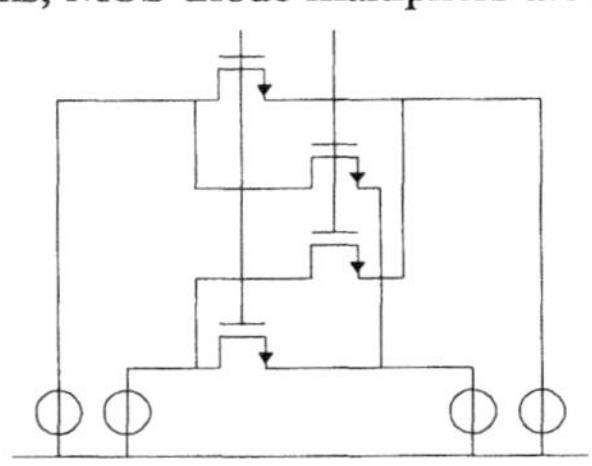

Fig. A12.8 MOS triode multiplier

[25] *Z.Czarnul, "Novel MOS resitive circuit for synthesis of fully integrated continuous-time filters", IEEE tr. Circuits and Systems, vol. 33, pp. 718-721, 1986*

[26] *Y.Tsividis, M.Banu, and J.Khoury, "Continuous-time MOSFET-C filters in VLSI", IEEE tr. Circuits and Systems, vol. 33, pp. 125-140, 1986*

[27] *S.Bibyk and M.Ismail, "Issues in Analog VLSI and MOS Techniques for Neural Networks", in Analog VLSI Implementation of Neural Systems, eds. C.Mead and M.Ismail, Kluwer, 1989, pp. 103-133*

[28] *F.M.A.Salam, M.R.Choi and Y.Wang, "Artificial Neural Nets in MOS Silicon", in Artificial Neural Networks and Statistical Pattern Recognition, eds. I.K.Sethi and A.K.Jain, 1991, pp. 243-270*

[29] *F.J.Kub, K.K.Moon, I.A.Mack and F.M.Long, "Programmable Analog Vector-Matrix Multipliers", IEEE JSSC, 1990, pp. 207-214*

[30] *T.Lehmann, "A Hardware efficient Cascadable Chip Set for ANN's with On-Chip Backpropagation", Int. journal of Neural Systems, vol. 4, pp. 351-358, 1993*

For proper operation, the multiplier must be driven and loaded with a low-ohmic load. For the drivers (the two leftmost voltage sources in figure A12.8) one typically uses OPAMPs. The lowohmic load (the two rightmost voltage sources) are either realized with OPAMPS or with current conveyers (CCII).

This circuit seems to be suited for application in on-chip learning neural networks. The input voltage range can be large, and the input signal voltage range does not matter. Problems may be in the construction of the driving voltage source and in the voltage-source load. Especially for high frequency operation, low impedant nodes are hard to realize. Lehmann reports settling times of 0.9μs.

Another multiplier that is based on MOS triode operation was publisher by Reed and Geiger. In this multiplier, the drain-source voltages of the triode-MOS transistors are set by the gate voltage of their cascode transistors.

[31] *R.D.Reed and R.L.Geiger, "A Multiple-Input OTA Circuit for Neural Networks", IEEE tr. CAS, 1989, pp. 767-769*

[32] *T.Kacprzak and K.Slot, "Multiple-input OTA Based Circuit for Cellular Neural Network Implementation in VLSI CMOS Technology", in Proc. CNNA 1992 , pp. 157-162*

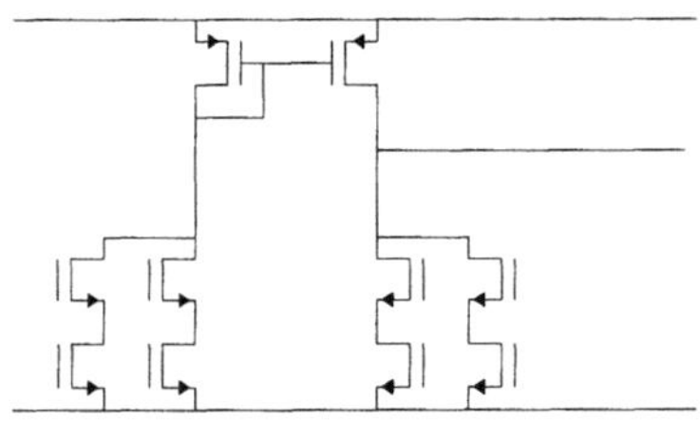

Fig. A12.9 Another MOS triode synapse

Linearized differential pairs

Linearized differential pairs have been published by among others Krummeracher. Vittoz proposed to use this multiplier in Kohonen networks. Although several methods to linearize differential pairs exist (including simple source degeneration), we show here only the one by Krummenacher. The main goal of linearizing differential pairs is evidently the increased linearity. For operation in neural networks, also the decreased transconductance is a major advantage as this decreases the accuracy demands on the weights (and on weight adaptation).

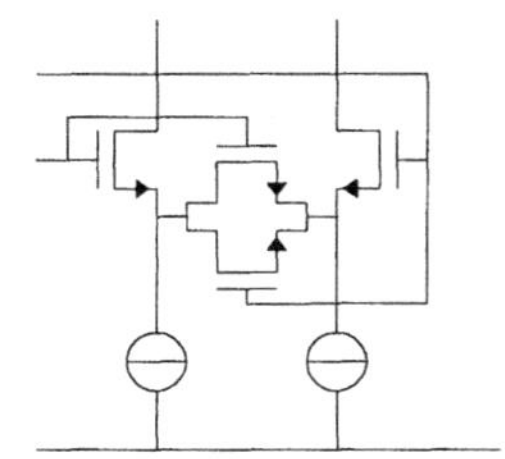

Fig. A12.10 Linearized differential pair

[33] *F.Krummeracher and N.Joehl, "A 4 MHz CMOS Continuous time filter with On-chip Automatic Tuning", IEEE JSSC, 1988, pp. 750-758*

Single inverters as synapses

Single inverters with grounded output nodes can be used as (fixed weight) multipliers. The weight is adjusted by the sizes of the MOS transistors in the multiplier; the circuit can therefore not be used in learning neural nets. Furthermore, literature indicates that non-adjustable neural networks are not likely to perform satisfactory on hard classification tasks because of the mismatch that occurs between the ideal neural net and the hardware implementation. Other disadvantages include the dependence on the supply voltage.

[34] *P.Masa, K.Hoen and H.Wallinga, "A High-Speed Analog Neural Processor", IEEE Micro, June 1994, pp. 40-50*

[35] *B.Nauta, "Analog CMOS Filters for Very High Frequencies", Boston: Kluwer Academic Press, 1993*

[36] *L.Tarassenko, J.Tombs and G.Cairns, "On-chip learning with Analogue VLSI Neural Networks", Int'l. Journal of Neural Systems, vol. 4, 1993, pp. 419-426*

Translinear multipliers

Although not very often applied in neural networks, translinear multipliers may be used for non-learning neural nets. For on-chip learning neural networks, the current inputs are a drawback because VI-conversion is required to input the weight voltage to translinear multipliers.

[37] *K.Bult and H.Wallinga, "A CMOS four-quadrant analog multiplier", IEEE J. Solid-State Circuits, vol. 21, pp. 430-435, 1986*

[38] *R.J.Wiegerink, "A CMOS Wideband Linear Current Attenuator with Electronically Variable Gain", on Proc. IEEE ISCAS, Chicago, pp. 962-965, 1993*

[39] *R.J.Wiegerinl, "Analysis and Synthesis of MOS Translinear Circuits", Boston: Kluwer, 1993*

Appendix 12.2: Neural activation functions: overview

This appendix presents an overview of literature that describes hardware realizations of activation functions for feed-forward neural networks is presented. Siliar to the previous appendix, this overview is not exhaustive but meant to give relatively fast a good overview. To limit the length of this appendix, related implementations are grouped together.

Differential pair-based activation functions
A large number of publications exist in which activation functions are implemented using CMOS differential pairs. The main differences between these implementations are:

- the type of mirror,
- including or excluding IV-conversion to obtain voltage output,
- including an OTA-like structure to increase the output voltage swing

In the three figures below, some realizations are presented. These realisations show the basic differential pair an some of the listed additions. Variable gain activation functions will be discussed later on in this appendix.

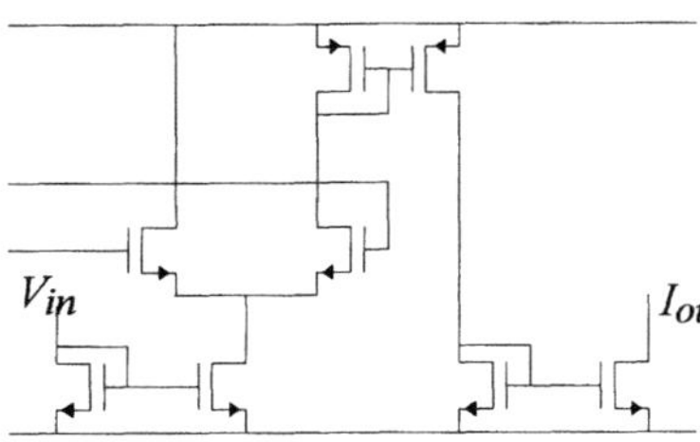

Fig. A12.11 Differential pair based activation function

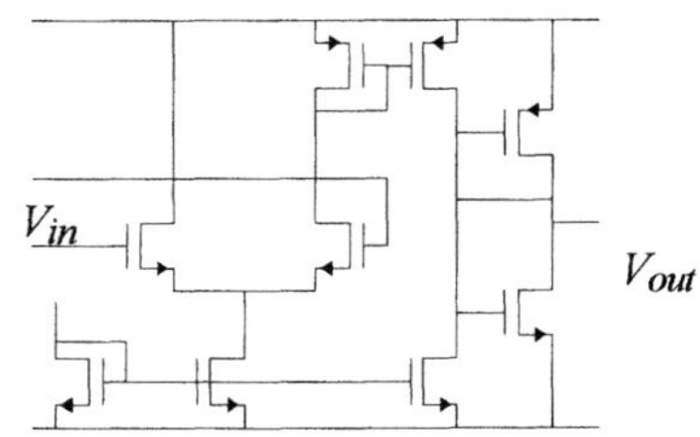

Fig. A12.12 Differential pair based activation function with voltage output

[1] *A.Nosratinia, M.Ahmadi and M.Shridhar, "Implementation Issues in a Multi-Stage Feed-Forward Analog Neural Network", IJCNN 1992, Baltimore, pp. II-642*

[2] *Lont and Guggenbuhl, "Analog CMOS Implementation of a Multilayer Perceptron with Nonlinear Synapses", IEEE tr. Neural Networks, 1992, pp. 457-465*

Variable gain, differential pair based activation functions

In the well known ETANN chip, a differential pair-based activation function with tunable gain is used. By tuning the gain, the slope of the non-linear function can be varied between almost linear to almost stepping. The circuit is shown below.

The shape of the sigmoid can be changed by $V_{control}$, which voltage controls the amplification of the input voltage V_{in}.

The ETANN chip is designed to be used for off-chip learning using a host computer. To minimize the effects of mismatch between the programmed neural network model and the actual implemented neural network, "chip-in-the-loop" training is used.

Because off-chip learning is used for the ETANN chip, no derivative approximating circuitry is included on-chip.

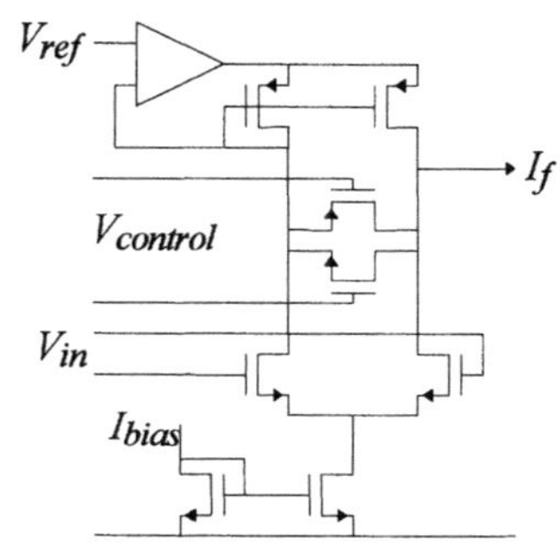

Fig. A12.13 Activation function with tunable slope

[3] *H.A.Castro, S.M.Tam and M.A.Holler, "Implementation and Performance of an Analog Nonvolatile Neural Network", Analog Integrated Circuits and Signal Processing, 4, 1993*

Duong et al. also presented non-linear functions with variable gain. These circuits are based on the replication and conversion into a symmetric current of the input current I_{in}. The output resistances of MOS transistors are then used to convert this current into a voltage. Gain control is included by the rightmost two current mirrors with forced control current.

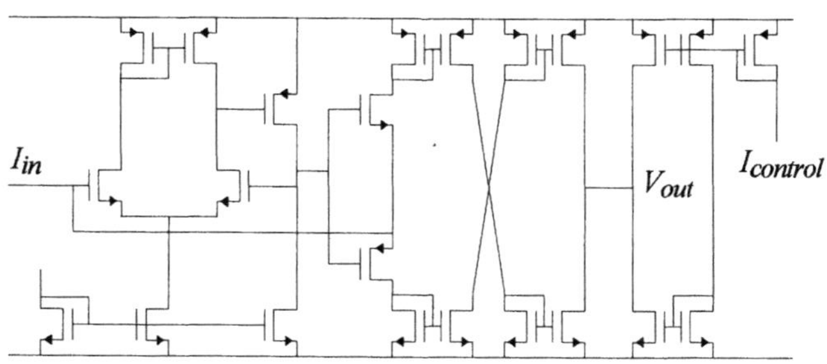

Fig. A12.14 Activation function with variable gain: according to Duong

[4] *Duong, Eberhardt, Tran, Daud and Thakoor, "Learning and Optimization with Cascaded VLSI Neural Networks Building-Block Chips", IJCNN 1992, Baltimore, pp. I-184*

Cascaded inverter-based activationa and derivative

Some attempts have been made to implement multi-layer on-chip learning neural networks with inverter-based activation functions. The circuit schematics of the activation function is shown in figure A12.5. In this circuit, the open-loop gain is large, and the actual gain of the block is fixed by a feed-back resistor. In order to match the output voltage swing of the non-linear function block and the input voltage swing of connected multiplier blocks, the output voltage swing is limited by the two voltage sources in the circuit. The resulting total output voltage swing is reported to be about 200mV.

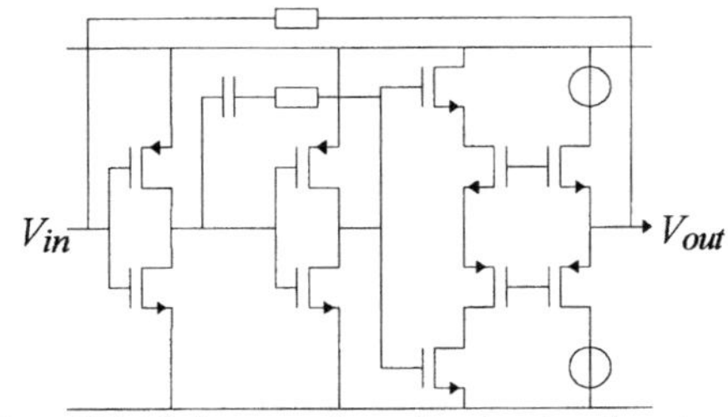

Fig. A12.15 Inverter-based activation function

Because this circuit is intended to be used in multi-layer on-chip learning neural nets, a derivative circuit is included. The next figure shows the total non-linear function block including the part that is used to approximate the first derivative. The basic idea behind this circuit is that the first derivative of a true sigmoid is given by $f'(x)=f(x)*(1-f(x))$. In the circuit, the actual non-linear function is not a true sigmoid, and hence the approximation of the derivative is not perfect.

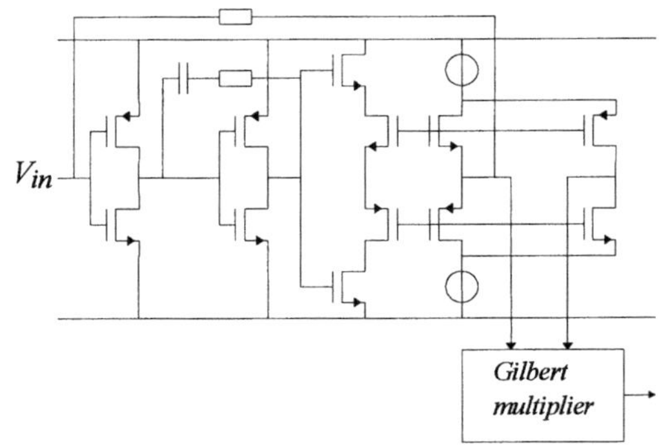

Fig. A12.16 Inverter-based Activation function and derivative approximation

[5] M.Valle, D.D.Caviglia, and G.M.Bisio, "An Experimental Analog VLSI Neural Chip with On-Chip Back-Propagation Learning", in Proc. ESSIRC 1992, Copenhagen, pp. 203-206

[6] D.D. Caviglia, M.Valle and G.M.Bisio, "A VLSI Module for Analog Adaptive Neural Architectures", in VLSI 91, eds. Halaas and Denyer, 1992

Inverter-based circuits

Several authors report on using cascaded inverters as activation functions in feed-forward neural networks. Choi presented cascaded inverters with an input VI converter to construct a sigmoid function. The circuit is shown below. The five leftmost transistors are used to convert the input voltage into a current. This current is then used to drive two cascaded inverters with feed-back resistances. The usage of MOS transistors for the resistances enables variable gain control.

[7] *Choi and Sheu, "VLSI Design of Compact and High-Precision Analog Neural Network Processors", IJCNN 1992, Baltimore, pp. II-637*

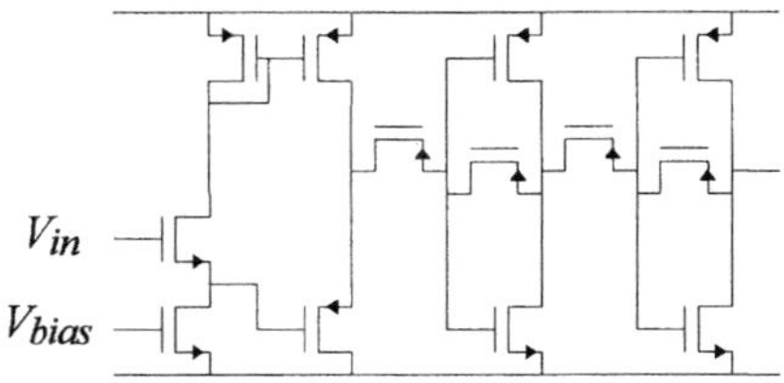

Fig. A12.17 Inverter-based activation function

Another variable-gain inverter-based activation function was published by Foo. This circuit is shown in figure A12.18. In the circuit, the non-linear function is constructed by the two cascaded inverters. The gain of these two inverters has been decreased by the two feed-back resistors. To include variable gain, a differential pair has been added.

[8] *S.Y.Foo. L.R.Anderson and Y.Takefuli, "Analog Components for the VLSI of Neural Networks", in IEEE Circuits and Devices, 1990, pp. 18-26*

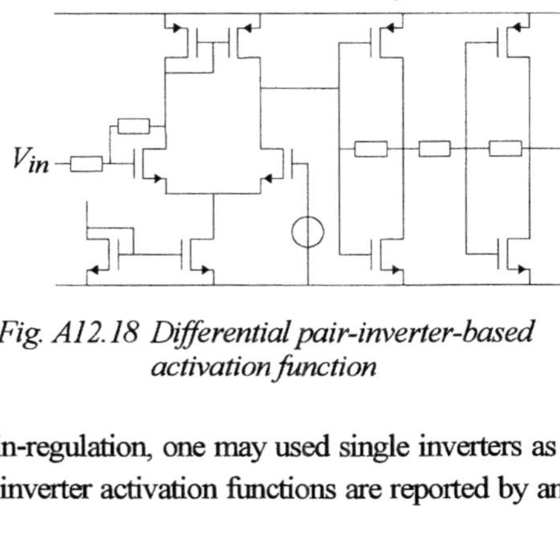

Fig. A12.18 Differential pair-inverter-based activation function

Instead of using cascaded inverters with gain-regulation, one may used single inverters as activation functions in neural networks. Single inverter activation functions are reported by among others

[9] *P.Masa, K.Hoen and H.Wallinga, "20 Million Patterns Per Second Analog CMOS Neural Network Pattern Classifier", in Proc. ECCTD 1993, Davos, pp. 497-502*

[10] *K.A.Boahen, P.O.Pouliquen, A.G.Andreou, R.E.Jenkins, "A Heteroassociative Memory Using Current-Mode MOS Analog VLSI Circuits", IEEE CAS, 1989*

[11] *A.Achyuthan and M.I.Elmasry, "A Methodology for Selection of Analog Hardware for Back-Propagation Algorithm Implementation", in Proc. IJCNN 1992, Baltimore, vol. 2, pp.672-677*

A hardware gaussian neuron

Watkins et al. presented at the IJCNN 1992 a design for a hardware realisation for a gaussian non-linear function. This non-linear function was intended to be used in radial basis function networks. This circuit is based on the selection of the minimum of the output currents of two differential pairs with a specific tail current relation. The resulting output current is gaussian-like with adjustable width and height. This function can be used as a radial-basis non-linear function.

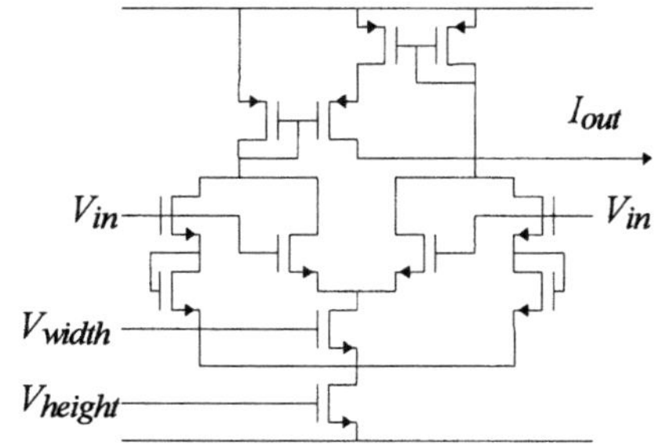

Fig. A12.19 A gaussian activation function

[12] S.S.Watkins, P.M.Chau and R.Tawel, "A radial Basis Function Neurocomputer Implemented with Analog VLSI Circuits", IJCNN 1992, Baltimore, pp. II-607

Differential pair based activation functions with derivative

There are very few differential-pair-based realisations of activation functions with their derivative. Some circuit implementations use a multiplier to construct an approximation of the derivative using the relation $f'(x) = f(x)\,(1 - f(x))$, which is valid for sigmoids only. Recently, Bogason implemented a type of numerical derivation in analog hardware. In his circuit (see figure A12.20) two differential pairs are used in parallel, one having the exact input voltage and the other having as input the sum of the original input voltage and a small differential voltage. The derivative is now derived by subtracting the outputs of both differential pairs, resulting in

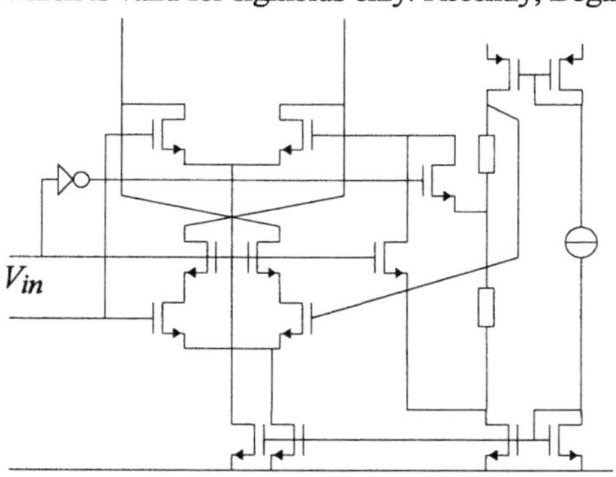

Fig. A12.20 Activation function and derivative (Bogason)

$$I_{out,der} \approx C\,\frac{f(x) - f(x+\delta)}{\delta}$$

[13] G.Bogason, "Generation of a neuron transfer function and its derivatives", Electronics Letters, 1993, pp. 1867-1869

[14] T.Shima, T.Kimura, Y.Kamatani, T.Itakura, Y.Fujita, and T.Iida, "Neuro Chips with On-Chip Back-Propagation and/or Hebbian Learning", IEEE J. Solid-State Circuits 1992, pp. 1868-1875

Chapter 13

Analog weight adaptation hardware

13.1 Introduction

In the previous chapter, some circuits that can be used in the forward-part of analog on-chip learning neural networks were presented. For on-chip learning neural networks, also on-chip weight adaptation hardware is required. In this chapter, a circuit for weight adaptation in on-chip learning feed-forward neural nets is described. This circuit must be very accurate because, as shown in chapter 6, the demands on the maximum parasitic charge injection during adaptation are hard to satisfy. Other demands on the weight adaptation circuit include accurate weight adaptation and a small chip area. Furthermore, the circuit must be fast enough to insure an acceptable speed of operation. The design of the weight adaptation circuit is presented in this chapter.

13.2 Multiplier: the basic idea

For charging and discharging of the weight storage capacitors using analog circuitry, a multiplier with current output is required. This current output must be gated to the weight storage capacitor during short intervals, denoted with phase θ, and the total charge that flows during θ is small. In order to keep acceptable gate-intervals, the output current of the weight adaptation circuit must be in the low nA range. This output current is the product of the input signal and the feed-back error signal, which both are in the current domain. The current multiplication can be performed by a CMOS differential pair in weak inversion in the configuration shown in figure 13.1. An MOS-switch gates the output current of this differential pair to the weight storage capacitor.

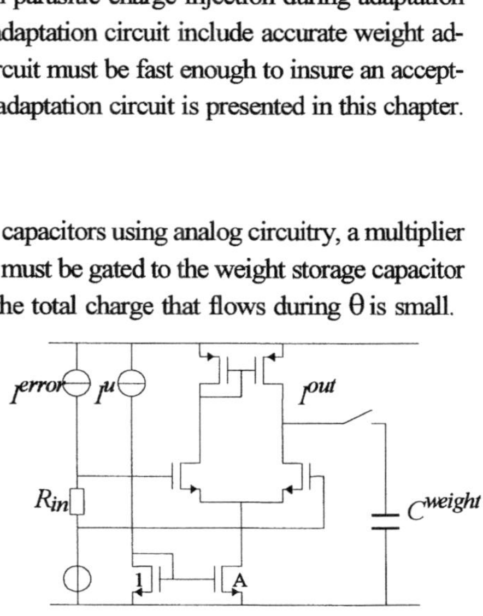

Fig. 13.1 Feedback multiplier: basic idea

In case of a closed switch, the output current of this multiplier is given by

$$I^{out} = I^{u} A \tanh\left\{\frac{e\, R_{in}\, I^{error}}{n\, k\, T}\right\} \tag{13.1}$$

$$n = \frac{C_{ox} + C_{depl}}{C_{ox}} \qquad 1.3 < n < 2.0$$

With the Taylor series expansion of tanh(x), it follows that for small input voltages the output current can be approximated by

$$I^{out} = I^{error}\, I^{u}\, A \frac{e\, R_{in}}{n\, k\, T} \tag{13.2}$$

When closing the switch during an interval τ, the voltage across the weight storage capacitor is approximately increased by

$$\Delta V^{weight} = \tau\, I^{error}\, I^{u} \frac{A\, e\, R_{in}}{n\, k\, T\, C^{weight}}$$

The equations presented in the previous section are derived for ideal MOS transistors and for a circuit that has no supply voltage limitations. In hardware realizations, however, a number of non-idealities prevent the circuit in figure 13.1 from operating properly.

Parasitic capacitance at the output node

At the output node of the differential pair is parasitic capacitance. In phase $\bar{\theta}$, i.e. in the phases in which the switch is opened, the output current of the differential pair either charges or discharges this parasitic capacitance. The voltage at the output node will approach respectively the supply voltage or the source voltage of the differential pair. The voltage increment on the storage capacitor due to only the parasitic capacitance at the output node is

$$\Delta V^{weight} \approx \begin{cases} \dfrac{C^{par}}{C^{par} + C^{weight}}\left(V_{dd} - V^{weight}\right) & \text{for } I^{error}\, I^{u} > 0 \\ \dfrac{C^{par}}{C^{par} + C^{weight}}\left(V_{bias} - V_{T} - V^{weight}\right) & \text{for } I^{error}\, I^{u} < 0 \end{cases}$$

Denoting the probability on a training example with $I^{error}\, I^{u} > 0$ that results in adaptation as P_p and denoting the probability on examples that result in adaptation with $I^{error}\, I^{u} < 0$ as P_n, the increment of the voltage across the weight storage capacitor due to charging and discharging of the parasitic capacitor at the output node of the differential pair is

$$\overline{\Delta V^{weight}} \approx \frac{C^{par}}{C^{par} + C^{weight}}\left[P_p\left(V_{dd} - V^{weight}\right) + P_n\left(V_{bias} - V_T - V^{weight}\right)\right] \tag{13.3}$$

This offset charge current is therefore decreasing during training due to decreasing probability on an example that results in adaptation. Furthermore, the offset denoted by (13.3) is heavily dependent on the stored weight voltage and supply voltage.

Mismatch

In hardware realizations, we have to consider V_T mismatch and geometry mismatch. For CMOS differential pairs operating in weak inversion and with an ideal current mirror, the effect of geometry mismatch is negligible. The offset input voltage is then by approximation equal to the V_T-difference.

For the current mirror, both V_T-mismatch and geometry-mismatch have an effect on the offset input voltage of the differential pair. The effect of the V_T-mismatch can be decreased by choosing a low W/L ratio and the effect of geometry mismatch can be decreased by decreasing the relative mismatch in geometry, i.e. by choosing large transistors.

For negligibly small effect of the offset voltage of the total differential pair, the offset voltage must be much smaller than the input voltage corresponding to a typical weight change. For the differential pair in weak inversion, the input voltage corresponding to a typical weight change will be well below $^{kT}\!/_{e}$.

Finite output resistance

A MOS transistor has a finite output resistance. Noting that the voltage across the weight-storage capacitor is generally not equal to the gate voltage of the PMOS transistors, it may be clear that finite output resistance of MOS transistors results in an offset output current of the differential pair. This offset output current is equivalent to an offset input voltage which depends on the voltage across the weight storage capacitor.

Leaky switch

As calculated in chapter 6, the leakage current for weight-storage capacitors must be very small. It is therefore important to decrease leakage through the switch. In hardware realizations, the switch generally consists of one or two MOS-transistors. The sub-threshold current can be decreased by both applying a gate-source voltage well below the threshold voltage or by applying a very low drain-source voltage.

When implementing the switches as MOS transistors, the leakage current results through the switch is mainly due to the current that flows through the reversed biased drain-bulk junction. To decrease this leakage current, the voltage across this junction must be made small (ideally zero).

13.3 Towards a solution

Figure 13.2 presents a circuit that eliminates most of the problems depicted in the previous section. The multiplication of two currents is performed by the leftmost differential pair, which is operated in weak-inversion. The rightmost differential pair is also operated in weak inversion, and serves as an offset-sensing and compensating circuit while keeping the voltage across switch S_{out} (ideally) at zero.

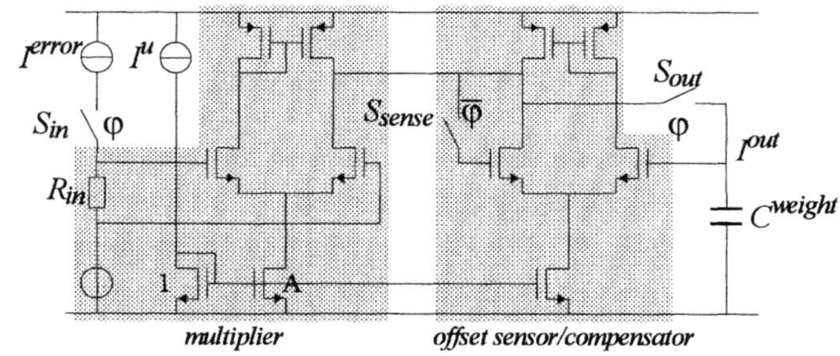

Fig. 13.2 Feedback multiplier: improved $I^{error}\, I^{u}\, \tau$ multiplier

The effects of charging and discharging of the parasitic capacitance at the output of the multiplier are significantly decreased due to the small voltage drop across switch S_{out}. This small voltage drop also reduces leakage through S_{out}.

Assuming that the offset current of both differential pairs is proportional to their source currents, the differential voltage at the input of the sampling-differential pair is quasi static during training and the (I_u dependent) offset currents of both differential pairs are compensated for by the rightmost differential pair.

Parasitic output capacitance

For the multiplier in figure 13.2, the effect of the parasitic capacitances is reduced because of the decreased voltage drop across these parasitic capacitances. The total parasitic capacitance is however increased due to applying two differential pairs in parallel. Denoting the input voltage required to compensate for the summed offset-output current as $V_{off,sum}$, the average increment of the weight voltage due to this offset voltage is

$$\overline{\Delta V^{weight}} \approx \{P_p + P_n\} \frac{C_{par}}{C^{par} + C^{weight}} V_{off,sum} \tag{13.4}$$

The differential input voltage of the offset-current sampling differential pair is generally larger than the offset voltage corresponding to the configuration of figure 13.1 because of operating two differential pairs in parallel. Furthermore, because mismatch is included in the offset voltage, it is still important to pay attention to minimize mismatch.

Parasitic output resistance

Finite output resistance of the two differential pairs will result in an offset output current. It can be derived that this offset output current is approximately

$$I_{offset} = V_{off,sum} \frac{R_{out1} + R_{out2}}{R_{out1}\, R_{out2}} \tag{13.5}$$

where: R_{out1} is the output resistance of the leftmost differential pair in figure 13.2

R_{out2} is the output resistance of the rightmost differential pair in figure 13.2

As the transconductance of a differential pair is usually much larger than the output conductance of a differential pair, (13.5) is usually much smaller than the compensated offset current. Note that as the output resistance of MOS transistors scales quadratic with the length and inversely proportional to the width, the output resistance can be increased at the cost of an increased chip area.

Drain-gate capacitance of offset-compensator

The drain-gate capacitance of the offset-compensation differential pair (see figure 13.2) degrades the circuit performance. On closing of S_{out}, the drain voltage of the leftmost transistor in the offset-compensator decreases $V_{off,sum}$. As a result of the presence of a drain-gate capacitor, the gate-source voltage of both transistors in the differential pair changes.

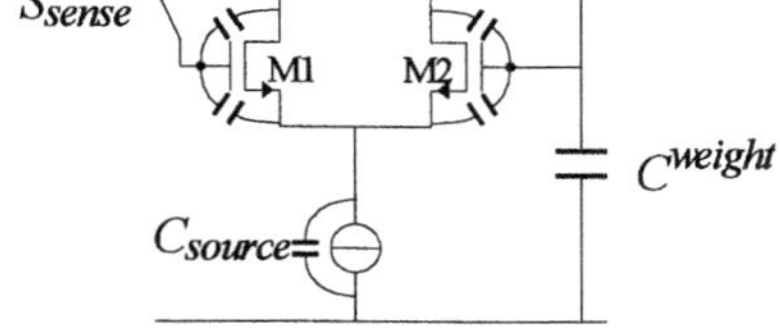

Fig. 13.3 Simplified equivalent circuit for effect of $C_{dg,1}$

When closing the S_{out}, the change in input voltage of the sampling differential pair is given by:

$$\Delta V_{in} = V_{off,sum} \frac{2\, C_{dg1}\, C_{gs} + C_{dg1}\, C_{source}}{C_{gs}\left(C_{dg1} + C_{gs} + C_{source}\right) + C_{dg1}\left(C_{gs} + C_{source}\right)} \tag{13.6}$$

This differential voltage results in an offset output current when S_{out} is closed. The average increment of the voltage across the weight storage capacitor due to this effect is

$$\overline{\Delta V^{weight}} \approx \frac{P_p + P_n}{C^{par} + C^{weight}} V_{off,sum} \frac{e\, \overline{I}^{u}}{n\, k\, T} \frac{2\, C_{dg1}\, C_{gs} + C_{dg1}\, C_{source}}{\left(C_{dg1} + C_{gs1}\right)\left(C_{gs2} + C_{source}\right) + C_{dg1}\, C_{gs1}} \tag{13.7}$$

Note that this current can be minimized by selecting smaller and longer transistors in the offset-compensation differential pair or by using a relatively large capacitor between the gates of the MOS transistors in the offset-sampler differential pair and ground.

Non-optimal switching

The clock signals φ and $\overline{\varphi}$ are ideally boolean inverse. With infinitely fast switches, this results in only surpassing both a pure offset-compensation phase and a pure weight-adaptation phase. In reality the clock signals have finite rise and fall times and the switches have a finite speed, which results in simultaneously be closed of S_{out} and S_{sense}. Because of the overlapping sense and multiply phases, the rightmost differential pair does not sample the offset output current, resulting in not-compensating the offset output current.

To insure that the offset-compensation differential pair operates properly, non-overlapping clock-pulses that realize a "break before make" behavior for S_{out} and S_{sense} must be used. However, in the period in which the circuit is switched in neither of the two modes, the parasitic capacitance at the output node C^{par} will be charged by the sum of the ideal output curren,the parasitic output current and the compensation current, see figure 13.4.

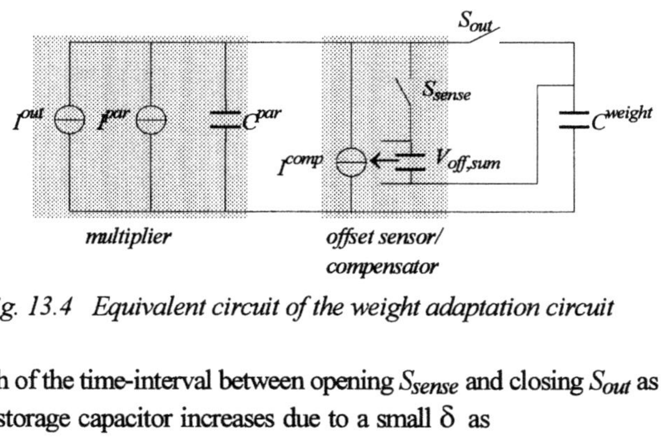

Fig. 13.4 Equivalent circuit of the weight adaptation circuit

Denoting the length of the time-interval between opening S_{sense} and closing S_{out} as δ, the voltage across the weight-storage capacitor increases due to a small δ as

$$\Delta V_{c,\delta} = \frac{(I^{out}+I^{par}+I^{comp})\,\delta}{C^{weight}+C^{par}} \tag{13.8}$$

Hence, a non-zero interval between φ and $\overline{\varphi}$ results in a virtually increased φ interval. Note that δ must be small in order to prevent the voltage across the parasitic capacitor from clipping.

13.4 The weight-update multiplier

Figure 13.5 shows the design of the weight-update multiplier. This multiplier consists of the previously analyzed CMOS differential pair in weak inversion to multiply I^{error} and I^{u}, and an offset compensator circuit which also consists of a weak inversion CMOS differential pair.

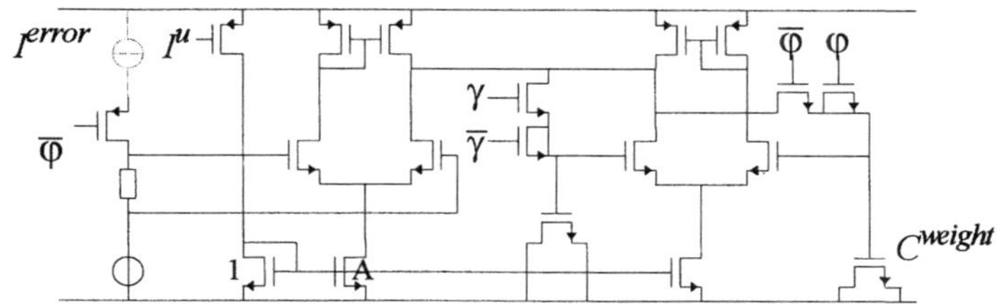

Fig. 13.5 Weight adaptation circuit

The S_{out} switch has been realized with NMOS transistors in a nulling scheme to decrease charge injection due to switching the MOS transistors [1]. The charge-injection at the output nodes of the differential pairs are not compensated in order to minimize parasitic output current (13.4). To minimize leakage when not-adapting, using PMOS transistors with their wells connected to the output node of the multiplier would be optimal. In that configuration, the voltage drop across the pn-junctions connected to the weight-storage capacitor equals the differential input voltage of the rightmost differential pair, which is usually small. As a result of this, the leakage current through the pn-junction connected to the weight storage capacitor would be small. However, using PMOS transistors for S_{out} increases the capacitance at the output node considerably which is undesired, see (13.4). Note that also when using an auxiliary voltage follower to copy the voltage at the output node to the well, the capacitance would increase significantly.

The S_{sense} switch is not critical with respect to leakage and is realized with NMOS transistors. To minimize the effect of the charge-injection at the gate of the leftmost transistor in the current sampling differential pair, an extra capacitor is used. Furthermore, the overlap between the offset-compensation phase and the charge-phase is omitted by using non-overlapping clock pulses γ and φ.

13.5 Simulation results

A number of simulation were done on a layout of the weight-adaptation-multiplier of figure 13.5 with minimum-sized switch-transistors, and assuming parameters of the UT-BICMOS process, which is the standard process at the University of Twente. The simulations of the weight-update multiplier show that the output current of the circuit (excluding the parasitic charge injection and leakage effects) agrees with (13.2). According to PSpice simulations, the constant leakage of the circuit is very small, resulting in a leakage current of

$$I^{leakage} \approx 10^{-12}\ A$$

This leakage can be decreased by a factor 3 by omitting the nulling MOS transistor in S_{out}. By using PMOS transistors with their bulk connected to the output node of the differential pairs, the leakage is decreased by about two orders of magnitude with respect to the circuit with NMOS transistors in a nulling scheme. Both methods can however not be applied in learning neural networks that use back-propagation: the omission of the nulling NMOS transistor results in large charge injection during switching and using PMOS transistors with their bulks connected to the output node of the multiplier increases the output capacitance considerable, which in turn increases the parasitic charge-injection significantly, see (13.4).

The charge injection due to switching the MOS transistors and due to the differential input voltage of the sampling differential pair is not negligibly small. According to PSpice simulations, the average parasitic charge injection on adaptation on one training example is

$$\Delta Q_{offset,sim,min} \approx 3.0 \cdot 10^{-15}\ \mathrm{C}.$$

This parasitic charge injection is mainly due to the charge injected when switching MOS transistors; the parasitic charge resulting from the offsets in the multiplier is lower by about a factor 5. With (6.18), the weight storage capacitor should *worst-case* be about 2–16nF for neural nets using back-propagation, with sigmoids having a *unity maximum derivative*, *double threshold* at 0.95 and 0.05, $\hat{V}^{weight}=1.5$; $\hat{W}=10$; η=0.1 and N_{in}=5. These values for the weight storage capacitor are unacceptably large for integrated capacitors.

When incorporating a regulator loop to further minimize the parasitic charge injection, one may gain one or two orders of magnitude in accuracy of the weight adaptation circuitry, and hence gain one to two orders of magnitude for the size of the weight storage capacitor. In the next section, a possible solution is presented and in the next chapter, a few other learning rules are proposed or discussed which may overcome accuracy demands to a certain extend.

13.6 Further reduction of charge injection

As shown in chapters 6 and 7 and in the previous sections of this chapter, the attainable accuracy in the weight adaptations is a major problem in analog hardwired neural networks. The (average) offset component of the weight adaptation must be very small in order to ensure proper convergence during training. In this section, a method initiated by Eichenberger and Guggenbuhl [1] and presented by Espejo et al. [2] is used to minimize the parasitic charge injection of the circuit presented in the previous section.

In analog neural networks, the weights are usually stored as voltages across capacitors; weight adaptation is then performed by injection charge onto the weight storage capacitor. In the circuit presented in the previous section, the parasitic charge injection due to switching MOS transistors is reduced by using dummy switches as proposed in for example [3]. In ordinary dummy switch techniques, the dummy switches are usually half the size of the actual switch-transistor which ideally compensates for the switch-induced charge injection.

Eichenberger and Guggenbuhl [1] presented an analysis of the effects of the load and drive and of the switching signals of the compensated switch. It followed that in most cases, simultaneously switching of the actual and of the dummy switch is not optimum with respect to cancellation of charge injection. Espejo et al. [2] presented a technique, based on the analysis of Eichenberger and Guggenbuhl, that can be used to adaptively change the switch-signals in such a way that charge-injection is canceled up to (reported) about 99.9%. This method is based on tuning of the timing between switching off of the actual switch and "turning on" of the dummy-switch. The charge injection and distribution can be changed by changing the phase between the two signals.

Charge injection cancellation in analog neural networks

The method of Espejo et al. was originally intended to be used for switched-current circuits. This section presents a short overview of the method of Espejo, when applied to weight adaptation circuits in analog neural networks. The block-schematic of the weight adaptation circuit including the dummy switch is presented in figure 13.6.

The current source I_{out} produces the load-current which charges the weight storage capacitor C^{weight} during φ; the output capacitance of the current source is denoted as C^{par}. To minimize leakage through the switches and to minimize the effects of C^{par}, the switched voltage follower is included in the weight adaptation circuit presented in the previous section of this chapter.

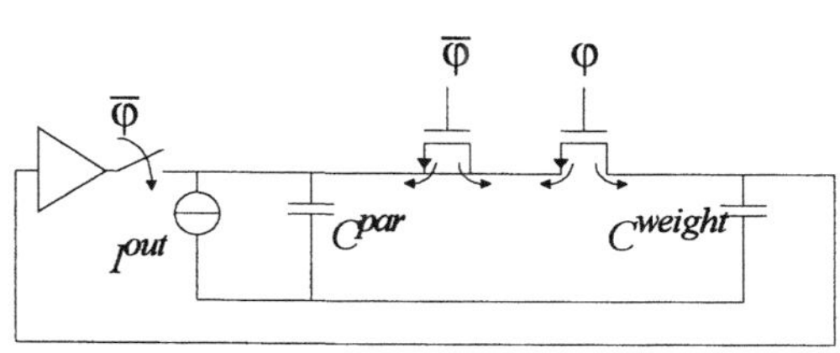

Fig. 13.6 Equivalent of weight adaptation circuit

Switching on: simple case

Assuming that the dummy switch is not present, the equivalent of switching-on the MOS transistor is shown in figure 13.7a. Due to switching, charge Q flows out of the drain and source terminals and the transistor is switched in triode region.

Because of charge redistribution between the parasitic capacitor and the weight storage capacitor, the net charge that flows into C^{weight} due to only switching is

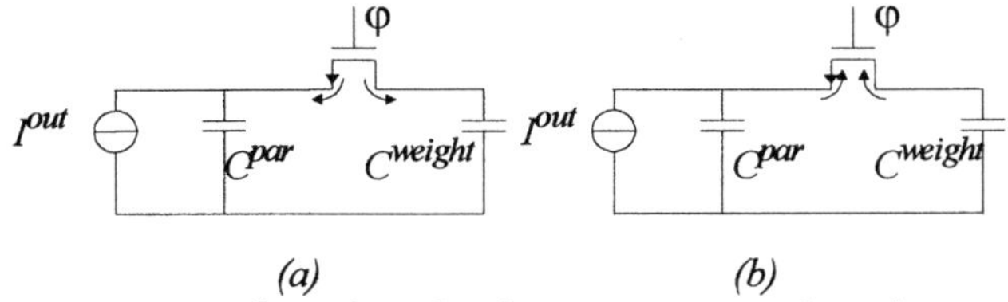

Fig. 13.7 Equivalent of weight adaptation circuit without dummy-switch: (a) switching on and (b) switching off

$$\Delta Q_{weight} = \frac{C^{weight}}{C^{weight} + C^{par}} \, 2\,Q$$

In general for the two capacitors $C^{weight} \gg C^{par}$, and hence virtually all charge flowing out of both the drain and source flows out of the weight storage capacitor.

Switching off: simple case

The situation in figure 13.7b depicts the switching-off of the MOS switch. Two fundamentally different situation may occur. Firstly, in case of *slow switching off* of the MOS transistor, the charge flowing into the drain and source can redistribute over the two capacitances because the MOS switch is not closed before redistribution has finished. In this case,

$$\Delta Q_{weight} = \frac{C^{weight}}{C^{weight} + C^{par}} \, 2\,Q\,.$$

Secondly, in case of *fast switching off* of the MOS switch, the charge flowing into the source and drain of the transistor can not redistribute because directly after injection of the charge, the MOS transistor is cut off. In this case,

$$\Delta Q_{weight} = Q$$

which is by approximation half the value in the previous case. In both cases, the charge injected on the C^{par} is absorbed by the voltage buffer which is switched on as soon as the MOS transistor is switched off.

It can be concluded that the charge injection cancels in first order when switching the MOS transistor slowly. However, when switching fast, the charge cannot redistribute and only half of the charge injection cancels in first order. To decrease the effects of the charge injection and to decrease the effects of the switching signals, one usually uses dummy switches that ideally absorb all injected charge of the MOS switch. Analyses in [1] show that for total absorption the switching signal of the actual switch and of the dummy switch must not be boolean inverse, but is ideally a "delayed boolean inverse".

Charge injection cancellation according to Espejo

To decrease the effects of charge injection, Espejo et al propose to use the method presented below. The dummy switch and the actual MOS switch have equal size.

During switching on of the MOS switch, the dummy switch ideally absorbs all injected charge of the MOS switch. Note that this is due to the charge redistribution of all injected charge over the two capacitors. Furthermore, the timing between φ and $\overline{\varphi}$ is not critical; the only condition being that the voltage buffer in figure 13.7 is disconnected during switching of either MOS transistor.

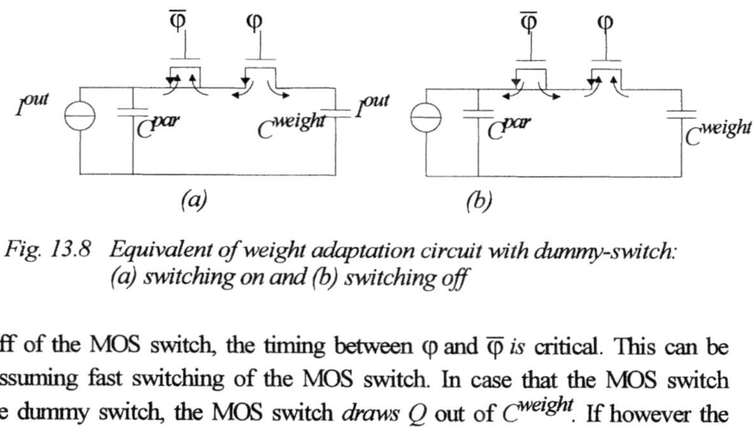

Fig. 13.8 Equivalent of weight adaptation circuit with dummy-switch: (a) switching on and (b) switching off

When switching off of the MOS switch, the timing between φ and $\overline{\varphi}$ *is* critical. This can be illustrated when assuming fast switching of the MOS switch. In case that the MOS switch switches before the dummy switch, the MOS switch *draws* Q out of C^{weight}. If however the dummy switches before the actual switch, the total *injected charge is* Q. By changing the phase between the falling edge of φ and the rising edge of $\overline{\varphi}$, one can modify the total injected charge onto C^{weight} in the range $[-Q, Q]$.

In the next part of this section, a closed loop system that may be used to cancel all remaining offset charges injected on the weight storage capacitor during adaptation (including offsets of the multiplier block) is presented. The system is however based on matching between transistors.

Charge-cancellation system
In this section a system which adaptively cancels charge injection and which is intended to be used in analog on-chip learning neural networks is presented. The system is shown in figure 13.8.

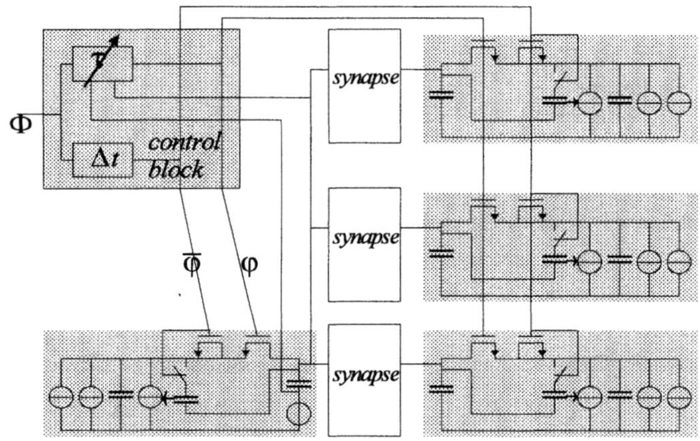

Fig. 13.9 Weight adaptation system for a neural network

In the system, one "dummy weight adaptation" block is used in the control loop. In this loop, the delay between the falling edge of φ and the rising edge of $\overline{\varphi}$ is adjusted in such a way that the voltage drop over the capacitor in the "dummy weight adaptation block" is stationary. It may be clear that the weight adaptation block in the dummy circuit must ideally have zero output current. In case that the voltage across the storage capacitor in the dummy block is stationary, the net charge-injection is zero.
To further decrease the effect of the charge injection on the synapses' outputs, the voltage at the non virtually-grounded node of the storage capacitor in the dummy circuit is used as reference voltage in the synapses. In this case, the sensitivity of the neural network with respect to errors in the control loop is decreased significantly.
Note that with good matching between the "dummy block" and the actual weight adaptation blocks, not only charge injection due to switching the MOS switches is eliminated, but *all offsets* of the weight adaptation multiplier are eliminated.

Matching between the blocks

In the charge-injection cancellation system, the matching between the dummy weight adaptation block and the normal weight adaptation blocks must be good. This means that no minimum sized transistors should be used; Espejo et al. report good results using $10\mu m^2$ transistors. Another point that causes mismatch between the dummy block and the normal weight adaptation blocks is the difference between the weight voltage and the reference voltage. It is however straight forward to show that this particular mismatch effect causes a shift between the effective clock signals φ and $\overline{\varphi}$. As a result of this phase shift, a type of bipolar weight decay (weight decay towards zero weights) [4] results. This type of weight decay is shown to be much less harmful for learning than unipolar weight decay (i.e. towards an extreme weight value). This type of "weight decay" is minimized by using very steep edge for the clock pulses.

The control block

In the control block, the timing of the falling edge of φ and the riding edge of $\overline{\varphi}$ is controlled. Espejo et al. propose to use a simple circuit which delays the falling edge by limiting the current that charges the input node of an inverter. In the control loop for the neural networks, the circuit would be something as shown in figure 13.10.

Using this circuit, the control loop adjusts the delay in such a way that the voltage across the dummy weight storage capacitor is stationary. Small fluctuations do not affect the network's operation in first order because of the symmetrical configuration.

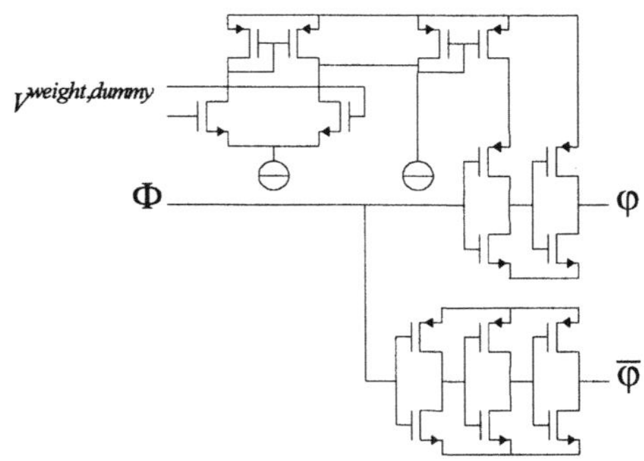

Fig. 13.10 Control block circuitry

13.7 Conclusions

It followed from the analyses in chapter 6 that the demands on (constant type) parasitic weight adaptation are very high: a small inaccuracy results in very large demands on the size of the weight storage capacitor in order to learn properly. In this chapter, a circuit that can perform analog weight adaptation in feed-forward neural networks is presented. In order to minimize parasitic weight adaptation, first order and second order techniques were applied. However, it was shown that even with these offset minimization techniques, the requirements on the parasitic weight adaptation are hard to meet. As a result of this, large weight storage capacitors would be required.

However, with a regulator loop as presented in section 13.6, it is expected that the parasitic weight adaptation of the circuit presented in this chapter can be reduced significantly. Other advantages of this regulator loop is that it results in a type of weight decay towards zero weights, which is reported to be much less harmful for the learning behavior of feed-forward neural nets. It follows that for medium and large sized training sets (not small N_{in}) of medium or high complexity (not small required $\hat{W}$) a regulator loop must be included to cancel weight-adaptation offsets. For too high weight adaptation offsets, the learning capabilities of the neural network degrade significantly.

13.8 References

[1] C.Eichenberger and W. Guggenbuhl, "On Charge Injection in Analog MOS Switches and Dummy Switch Compensation Techniques", IEEE tr. Circuits and Systems, vol. 37, no. 2, pp. 256-264, 1990

[2] S.Espejo, A.Rodriguez-Vazquez, R.Dominguez-Castro, J.L.Huertas, "A Modified Dummy-Switch Technique for Tunable Feedthrough Cancellation in Switched-Current Circuits", ESSCIRC 1993, pp. 270-273

[3] R.Gregorian and G.C.Themes, "Analog MOS Integrated Circuits for Signal Processing", New York: Wiley, 1986

[4] D.B.Mundie and L.W.Massengill, "Weight Decay and Resolution Effects in Feedforward Artificial Neural Networks", IEEE tr. Neural Networks, pp. 168-170, 1991

Chapter 14

Conclusions

In this book, a method to mathematically analyze the learning behavior of (multi-layer) feed-forward neural networks, the Vector Decomposition Method (VDM), has been introduced. With the VDM, a large number of phenomena during training of feed-forward neural networks have been analyzed mathematically, which results in easy-to-read equations and hence results in insight into the learning processes in feed-forward neural networks.

In literature, most modifications to the neural network and its learning algorithm are intuitive. With the VDM, *motivated* modifications to both the network and to the adaptation algorithm can be made. These motivated modifications either simplify the implementation of the network (in analog hardware, digital hardware or software) or speed up the convergence during training significantly. In this book, the VDM has also been used to both derive specifications for fully analog hardware implemented on-chip learning feed-forward neural networks and to find simple (hardware implementable) modifications to the learning algorithm that result in faster learning. However, most of the analyses and conclusions can also be applied to feed-forward neural networks that are implemented in either software or digital hardware.

Section 14.1 summarizes the original contributions of this book, and section 14.2 presents some recommendations for further research.

14.1 Original contributions

The most important original contributions in this book are summarized below.

- A mathematical method for the analysis of the learning behavior of feed-forward neural networks (the Vector Decomposition Method, VDM) is introduced (chapter 2)
- The VDM has been used to derive easy-to-read equations that describe the learning behavior of single-layer and single-output two-layer neural networks (chapters 3 to 11)
- The VDM has been used to find easy-to-read relations between hardware-related problems of the feed-forward part of the network and the learning behavior of feed-forward neural networks (chapter 4, 5, 6, 7, 9 and 11)
- The VDM has been used to derive specifications for analog on-chip weight adaptation circuitry and has been used to simplify the hardware implementation while decreasing the demands on the building blocks (chapters 4, 5, 6, 9 and 11)
- In chapters 12 and 13, new circuitry to be used in on-chip learning neural networks has been presented.

14.2 Recommendations for further research

In this section, we will present a number of recommendation for further research. Where possible, a short discussion is given or a short introduction into the problem or into the solution is provided.

Analysis of more hardware-related problems

The VDM has been used to analyze the effects of a large number of parameters and hardware-related effects on training. The effects of especially a number of hardware related problems such as for example linearity of multipliers, noise effects, and resolution in multi-layer feed-forward neural networks has not been done yet. A mathematical analysis of these effects may result in additional specifications for building blocks. The analysis of resolution is merely an extension of the analyses in chapters 6 and 7, which could be done in the same way as the extension of chapter 4 in chapter 9:

When assuming that the gain of a two-layer feed-forward neural network is by approximation equally distributed over both layer, the i^{th}neuron in the first layer is activated in the middle region for (training) examples out of S_i (see also section 9.4). For these examples, the $\alpha_{1,i}^{h,v}$ that corresponds to a certain network response Y^p is given by

$$\alpha_{1,i}^{h,v} \approx \frac{f^{-1}(Y^p)}{\beta_{1,i}^{h} f'(0) \left(\underline{W}_{2,1}\right)_i NH_i}.$$

where: NH_i denoted the number of neurons classifying S_i

The situation that a cluster of neurons is used to classify a sub-set S_i will not be analyzed explicitly; unless stated explicitly, it will be assumed that sub-set S_i is classified by one (non-redundant) neuron in the first layer. It can however be shown that this situation is equivalent to a situation in which S_i is classified by NH_i neurons in the first layer with a different $\underline{W}_{2,1}$. The $\alpha_{2,1}^{h,v}$ is then equal to the $\alpha_{1,i}^{h,v}$ multiplied by the gain of the ith neuron in the first layer, $\alpha_{2,1}^{h,v} \approx \frac{f^{-1}(Y^P)}{\left(\underline{W}_{2,1}\right)_i}$.

A worst-case estimation of the maximum acceptable value of the constant weight adaptation is then given by:

$$\Delta W_{2,1}^{par} \leq \frac{\eta}{20\left(\underline{W}_{2,1}\right)_i} \cdot \int_{LinSep.} P(D-Y)\frac{(D-Y)f'\left(f^{-1}(Y)\right)f^{-1}(Y)}{P(correct)}\, d(D-Y).$$

The corresponding equation for the neurons in the first layer is then

$$\Delta W_{1,i}^{par} \leq \frac{\eta}{20\sqrt{N_{in}-1}\,\beta_{i,1}^{h,end}} \cdot \int_{Lin.Sep.} P(D-Y)\frac{(D-Y)f'\left(f^{-1}(Y)\right)f^{-1}(Y)}{P(correct)}\, d(D-Y).$$

From the last two equations, it follows that worst-case the demands on the parasitic constant weight adaptation are different by a factor $\sqrt{N_{in}-1}\,\beta_{1,i}^{h}/\left(\underline{W}_i\right)_i$. With the assumption that the gain of the neural network be equal for both layers, the ratio between the two tolerable constant weight adaptations is the square root of the weights in $\underline{W}_{1,i}$ and $\underline{W}_{2,1}$ that are responsible for the classification of sub-set S_i.

In back-propagation, the weight vector of *all* neurons in a neural network are modified if the response of the total neural network on any specific training vector is incorrect. However, the adaptation of $\underline{W}_{1,i}$ is ideally non-zero only for training examples in S^i, and is ideally zero for all other training examples:

$$\Delta W^{total} = \begin{cases} \Delta W^{ideal} + \Delta W^{par} & \text{for } \underline{U} \in S^i \\ \Delta W^{par} & \text{for } \underline{U} \notin S^i \end{cases}$$

Let us now define the probability that an training example lies in the sub-space S^i as $P(S^i)$ and let's define the probability on an example that lies in S^i results in any weight adaptation as $P(S^i \mid effective)$. The parasitic weight adaptation occurs during any weight adaptation cycle while the ideally non-zero weight adaptation occurs only for $P(S^i \mid effective)/P(S^{total} \mid effective)$ of all examples that result in weight adaptation.

When assigning all weight adaptation of the ith neuron in the first layer to training examples $\underline{U} \in S^i$, the equivalent weight adaptation of the ith neuron in the first layer of a two-layer neural network can now be written as

$$\Delta W_{1,i}^{total,equiv} \approx \begin{cases} \Delta W_{1,i}^{ideal} + \Delta W_{1,i}^{par}\dfrac{P(S^{tot} \mid effective)}{P(S^i \mid effective)} & \text{for } \underline{U} \in S^i \\ 0 & \text{for } \underline{U} \notin S^i \end{cases}.$$

The virtually enlarged effect of the parasitic weight adaptation is completely caused by the (ideally zero) weight adaptation of weights in $\underline{W}_{1,i}$ on training examples that are not in S^i. It may be clear that for two-layer neural networks, the demands on the accuracy for the weight adaptation circuitry increases linearly with the number of non-redundant neurons in the first layer (i.e. the number of distinct sub-classes S_n). A solution is straight forward.

Research into more accurate circuits
The weight-adaptation circuit in chapter 13 has parasitic charge-injection during adaptation which prevents neural networks with this circuit from learning high-dimensional training sets with some level of complexity. Research towards multipliers that have a lower parasitic charge injection could extend the applicability of fully analog on-chip learning neural networks.

Complexity reduction
It has been shown that fully-analog on-chip learning feed-forward neural network can be applied to low-dimensional training sets with limited complexity (i.e. limited weight range and limited network size). For high-dimensional training sets, a digital (based) weight adaptation circuit appears necessary. However, using for example a *principle component* network to reduce the dimension of the input vectors, it may be possible to apply fully analog on-chip learning neural networks with a pre-processor neural network to high dimensional training sets. Investigation of this type of hybrid neural networks seems useful.

Development of hardware dedicated algorithms
In this book, the learning algorithm is back-propagation. It is shown in this book that a large number of problems arise during training, especially for on-chip learning. Development of new hardware-friendly algorithms appears therefore to be very useful. We will discuss two:

In the back-propagation algorithm, the weight adaptation for any training example is small. In feed-forward neural nets that are implemented either in digital hardware or in software, this learning rate must be small to ensure convergence, i.e. to ensure that steps are small enough to find a minimum in the energy landscape. When implementing the back-propagation algorithm directly in analog hardware, the weight adaptation due to parasitic charge injection is large related to desired weight adaptations. This large parasitic charge injection prevents the neural network from learning properly.

In contrast to digital implementations of neural nets, the weight adaptation in continuous time analog hardware neural nets is allowed to be large, assuming that the weight change is of relatively low frequency for the analog circuit. In this case, the analog neural network converges towards a minimum in the energy landscape for each individual training example. This algorithm will be denoted as *continuous time back-propagation*. The weight adaptation of neurons in the second layer is then

$$\Delta \underline{W_2} = \eta \int_{t(0)}^{t(D=Y)} \left(D - Y(t)\right) f'\left(\underline{W_2}(t) \cdot \underline{U_2}(t)\right) \underline{U_2}(t)\, dt,$$

where $t(0)$ is the time at which weight adaptation on the training example starts, and where $t(D=Y)$ is the time at which the actual response of the neural net equals the desired response. The weight adaptation of the neurons in the first layer is

$$\Delta \underline{W}_{1,j} = \eta \int_{t(0)}^{t(D=Y)} \left(D - Y(t)\right) f'\left(\underline{W}_2(t) \cdot \underline{U}_2(t)\right) \underline{W}_2[j](t) f'\left(\underline{W}_{1,j}(t) \cdot \underline{U}_1\right) \underline{U}_1 \, dt,$$

The continuous time back-propagation algorithm has several advantages over the original back-propagation algorithm with respect to analog hardware implementations of neural nets. Firstly, the weight adaptation on training examples is large related to the original back-prop algorithm which reduces the *relative* effect of the charge-injection due to switching MOS switches. Secondly, the first derivative of the non-linear function of the neuron in the last layer can be omitted. This omission does not affect the weight adaptation on any example, but decreases the adaptation time $t(D{=}Y) - t(0)$ *and* reduces accuracy demands in the hardware net. Thirdly, the factor $\left(D - Y(t)\right)$ can be replaced by the sign of this factor, because this sign does not change during adaptation on one example and because the start and ending point of adaptation are not changed with this replacement. This replacement also decreases the adaptation time for one example and decreases accuracy demands because signal levels are closer to their maximum values.

To minimize parasitic charge injection due to switching MOS transistors, it appears to be useful to disable switches for weight updates if the derivative of the corresponding neuron in the first layer is below a certain threshold. In this way, switching that results in small weight adaptations are suppressed, which method has similarities with the trinary weight update algorithm of Shoemaker et al [4].

A trivial difference between both algorithms is that the continuous time version requires fewer training examples to reach a specific level of performance, which is due to the relatively large adaptation on each example. Simulations show that typical reduction of the number of training examples is about a factor 10. For non-linearly separable training sets, the continuous time learning algorithm causes oscillating hyperplanes, which is due to adapting until the desired response is obtained. According to simulations using the neural network simulation program NeuroSim, the maximum tolerable charge-injection for the continuous time back-propagation algorithm is about a factor 20 higher than for normal back-propagation, while the analog hardware implementation of the continuous time algorithm is considerably simpler than the implementation of the normal back-propagation algorithm.

Another (family of) weight adaptation algorithm [1] which is assumed to be hardware friendly is the weight perturbation algorithms and derivatives of weight perturbation. In the weight perturbation algorithm (and in its successors [2],), the weight adaptation needs only feed-forward circuits. A comparison of the network response and the network response after perturbing weight provides information about how to modify weights in order to minimize a predefined cost-function. The original formulation of weight perturbation was extremely slow because for all weights in a feed-forward neural network, weight perturbation was applied individually (sequentially). Recent parallel weight perturbation-based algorithms reduced the number of feed-forward calculations to only 2 for each input vector (back-propagation requires one feed-forward calculation and one feed-back calculation). Especially these recent parallel algorithms may become important for on-chip learning neural networks. It therefore appears useful to apply the VDM to this type of algorithms.

Using an other type of weight storage
The high accuracy demands in chapter 6 can significantly be decreased when using *two* weight storage elements instead of one single weight storage capacitor. In this case, the weight would be the sum of a non-volatile weight and an relatively small analog weight. During training, only the analog weight is adapted which decreases the demands on weight adaptation accuracy significantly, see section 6.7, because the weight range corresponding to this analog volatile weight is relatively small. Upon underflow or overflow of this analog weight, the non-volatile weight is updated and the analog weight is reset to zero. The non-volatile weight can be implemented either with EEPROM devices [5] or can be implemented using the synapse of figure A12.5 with an "analog least significant bit" [3].

Other algorithms
In this book, the VDM has been used for analyses of feed-forward neural networks that perform a pattern classification task. Extension of the analyses in this book towards feed-forward neural networks trained to perform a function approximation task is straight forward.
However, it is expected that the VDM can also be applied to the analysis of other neural network models such as self-organizing feature maps, cellular neural networks etc. Research into the wider applicability of the VDM is useful in order to derive insight into for example the learning processes of other types of neural networks. These analyses could be used to derive specifications for building blocks and to speed up learning by modifying either the learning algorithm or the network structure in such a way that the bottlenecks are omitted.

[1] M.Jabri and B.Flower, "Weight Perturbation: An optimal architecture and learning technique for analog VLSI feedforward and recurrent multi-layer networks", IEEE tr. Neural Networks, pp. 154-157, 1992

[2] M.Jabri, "Practical Performance and Credit Assignment Efficiency of Analog Multi-Layer Perceptron Perturbation Based Training Algorithms", MicroNeuro '94, Turin, 1994, Italy

[3] T.Lehmann, "Hardware Learning in Analogue VLSI Neural Networks", Ph.D. thesis, Technical University of Denmark, 1994

[4] P.A.Shoemaker, M.J.Carlin and R.L.Shimabukuro, "Back propagation Learning with Trinary Quantization of Weight Updates", Neural Networks, pp. 231-241, 1991

[5] G.v.Steenwijk, K.Hoen and H.Wallinga, "A Nonvolatile Analog Programmable Voltage Source Using the VIPMOS EEPROM Structure", IEEE J. Solid State Circuits, pp. 784-788, 1993

Index

Nomenclature

$\underline{B}_{k,j}^{bias}$:	unit vector corresponding to only the bias input of the jth neuron in the kth layer
$\underline{B}_{k,j}^{h}$:	unit vector perpendicular to the attractor hyperplane of the jth neuron in the kth layer
$dist$:	distance between two hyperplanes corresponding to neurons in the first layer (two neuron in the first layer assumed); to be used for translation-based breaking
D^p	desired (or target) response corresponding to the pth training example
E :	cost function to be minimized during training
MSE :	mean squared error of the neural network
$MSE_{min,r}$:	minimum mean squared error of the neural network with redundancy
$MSE_{min,nr}$:	minimum mean squared error of the neural network without redundancy
N_{crit} :	number of examples passed during training until β_{crit} is reached
t :	training time
$\underline{U}_k$:	input vector for the kth layer including the bias input signal
$\underline{U}_{k,j}^{CG.}$:	centre-of-gravity of the jth neuron in the kth layer
$\underline{U}_{k,j}^{F}$:	input vector component of the jth neuron in the kth layer in parallel to the associated attractor hyperplane.
$\underline{U}_{k,j}^{F,const}$:	(in a loose way) the constant part of $\underline{U}_{k,j}^{F}$ for examples that result in non-zero adaptation
$\underline{U}_{k,j}^{F,v}$:	(in a loose way) the zero mean part of $\underline{U}_{k,j}^{F}$ for examples that result in non-zero adaptation
$\underline{W}_{k,j}$:	weight vector of the jth neuron in the kth layer
$\underline{W}_{k,j}^{ATT}$:	weight vector corresponding to the so-called attractor hyperplane of the jth neuron in the kth layer
$\underline{W}_{k,j}^{\varepsilon}$:	weight vector component of the jth neuron in the kth layer in parallel to the associated attractor hyperplane.
$Y_{k,j}$:	output of the jth neuron in the kth layer
Y_{net} :	output of the (single output) network
$\alpha_{k,j}^{bias}$:	value of the bias element in the input vector of the the kth layer (identical for all j)
$\alpha_{k,j}^{h}$:	norm of the input vector of the the kth layer perpendicular to the attractor hyperplane of the jth neuron in the kth layer

$\alpha_{kj}^{h,\nu}$: (in a loose way) the zero-mean part of α_{kj}^{h} for examples that result in non-zero adaptation

α_{kj}^{ATT}: (in a loose way) the average part of α_{kj}^{h} of which the jth neuron in the kth layer learns

β_{crit} : value of β_{kj}^{h} for two initially redundant neurons for which redundancy starts to be lifted

β_{max} : value of β_{kj}^{h} for two initially redundant neurons after an infinite training in which the redundancy is not lifted

β_{kj}^{bias} : value of the weight connected to the bias input of the jth neuron in the kth layer

β_{kj}^{h} : norm of the weight vector of the jth neuron in the kth layer perpendicular to the associated attractor hyperplane

β_{valid} : value of β_{kj}^{h} for two initially redundant neurons for which the analysis conditions are marginally satisfied

η: learning rate

φ: angle between two hyperplanes corresponding to neurons in the first layer (two neuron in the first layer assumed); to be used for rotation-based breaking

$\Delta\beta_{1}^{bias}$: difference of two bias-related weights β_{1j}^{bias}(assuming two neurons in the first layer) to be used for translation-based breaking of clustered hyperplanes.

$\underline{\Delta W}^{\varepsilon}$: difference vector between two $\underline{W}_{1j}^{\varepsilon}$ vectors (assuming two neurons in the first layer)

τ_{crit} : training time t at which β_{crit} is reached

GPSR Compliance

The European Union's (EU) General Product Safety Regulation (GPSR) is a set of rules that requires consumer products to be safe and our obligations to ensure this.

If you have any concerns about our products, you can contact us on ProductSafety@springernature.com

In case Publisher is established outside the EU, the EU authorized representative is:

Springer Nature Customer Service Center GmbH
Europaplatz 3
69115 Heidelberg, Germany

Batch number: 10620141

Printed by Printforce, the Netherlands